EUR

The Pathological Protein

The Pathological Protein

Mad Cow, Chronic Wasting,
and Other Deadly Prion Diseases

Philip Yam

COPERNICUS BOOKS

An Imprint of Springer-Verlag

Published in the United States by Copernicus Books, an imprint of Springer-Verlag New York, Inc.
A member of BertelsmannSpringer Science+Business Media GmbH

Copernicus Books
37 East 7th Street
New York, NY 10003
www.copernicusbooks.com

Book design and line art by Jordan Rosenblum.

Library of Congress Cataloging-in-Publication Data
Yam, Philip.
  The pathological protein: mad cow, chronic wasting, and other deadly prion diseases / Philip Yam.
    p. cm.
  Includes bibliographical references and index.
  ISBN 0-387-95508-9 (alk. paper)
    1. Prion diseases—Popular works. I. Title.
  RA644.P93 Y35 2003
  616.8'3—dc21   2002042730

Manufactured in the United States of America.
Printed on acid-free paper.

9 8 7 6 5 4 3 2 1

ISBN 0-387-95508-9   SPIN 10880347

To my father, Peter, and my mother in memoriam, Wanda

CONTENTS

CHAPTER 9   Mad Cow's Human Toll

Figuring out how many people will succumb to variant Creutzfeldt-Jakob disease isn't easy—especially now that BSE has spread around the world.

CHAPTER 10   Keeping the Madness Out

Several measures help ensure that animal prion diseases do not contaminate the U.S. food supply—but there are gaps.

CHAPTER 11   Scourge of the Cervids

Chronic wasting disease of deer and elk, once confined to a patch in the Rockies, spreads across the nation.

CHAPTER 12   Misadventures in Medicine

Prion diseases spread to humans through medical mishaps.

Acknowledgments

This book is really a story of many career-long struggles to understand a rare family of neurodegenerative diseases, and I could not have told it without the time and attention that so many researchers granted me. Their contributions should be clear in these pages. My gratitude also goes to David and Dorothy Churchill, who allowed me into their home and re-lived their loss for my sake.

Friends and colleagues pitched in in many ways that made the book writing possible. John Rennie, Ricki Rusting, Gary Stix, and Molly Frances at *Scientific American* gave support, provided notes, reviewed drafts, and watched over my magazine duties while I worked on this project. I thank Bridget Gerety and Ed Bell for their help on the photographs, David Labrador for his fine-toothed fact-checking, and Julia Karow for her skilled translations of technical articles from old German journals. I am particularly indebted to Steve Mirsky, whose research assistance and draft comments were indispensable, and to Karen Hopkin, who helped get me started on this project and gave expert insight, guidance, and input in several areas of molecular biology. Additional valuable feedback came from Michael Yam and Lucy Chou, who also cheerfully did the driving on the opposite side of the road while we were in the U.K.

I thank my agent, Jennifer Gates of the Zachary Shuster Harmsworth Agency, editors Tim Yohn and Lyman Lyons, and designers Jordan Rosenblum and Stephanie Blumenthal. At Copernicus Books, thanks to Anna Painter, Mareike Paessler, and especially Paul Farrell, whose commitment and belief in the project saw it through some trying times.

Philip Yam
March 2003

Introduction

As I reclined in the patient chair in my dentist's Midtown Manhattan office last February, waiting for a lead apron and the film for bitewing x-rays, I noticed the stainless steel instruments glistening under the hard light of the examination lamp. Sharp hooks and pointed probes were carefully laid out on the tray. Some were still in their plastic wrapping, indicating that they had been sterilized. Soon, the dentist would push these tools between my teeth, under my gums, down the little pits in my enamel. One of the instruments might even draw a bit of blood — not an uncommon occurrence when the gums are inflamed by plaque buildup and sharp bits of metal are involved.

Then a thought occurred to me. "Do you have British patients?"

"Oh sure. I had two from Europe the other day," my dental hygienist replied as she circled behind the chair to clasp a paper bib around my neck. A lot of transatlantic traffic comes through this New York City office, she added. Some Europeans, she guessed, prefer the American approach of preventive dental care.

And that's when I realized that my risk of catching the human form of mad cow disease from these instruments was not zero.

You might wonder: How could this be? Are the instruments not sterilized? The answer is yes — and no. Many surgical and dental tools are steam-heated for 15 to 30 minutes at some 121° C (250° F). These scorching temperatures are more than a match for the bacterium that causes tuberculosis and the viruses that cause AIDS and hepatitis. In fact, you name it, and time and temperature in the autoclave will take care of it. Yet such extreme conditions cannot completely destroy the "mad cow" agent that, over time, peppers the brain with microscopic holes, causing clumsiness, dementia, and eventually death. Even formaldehyde, which can kill germs as well as human cells, does nothing to this microscopic killer.

Because this powerful agent resists standard sterilization, it can be spread from person to person through medical and, in principle, dental procedures. Although such documented cases involved procedures that were much more invasive than teeth cleaning (such as the implantation of electrodes into the brain), animal studies suggest that the illness can be transmitted via dentistry. And because the British population has been exposed to mad cow disease and no tests exist to determine who might be incubating the neuron-destroying agent, it is theoretically possible that the pathogen lurks on the scrapers, picks, probes, and other tools in my dentist's office.

Death from the Pathological Protein

The brain-eating invader is thought to be a protein called a prion—a *pro*teinaceous *in*fectious particle, as named by Nobelist Stanley Prusiner in 1982. Strictly speaking, the acronym should be "proin"—but in a savvy marketing move, Prusiner wisely transposed the "o" and the "i," lest people die from "proin" injuries. He also chose to pronounce prion as "*pree*-on," rather than the more phonetically natural "*pry*-on."

As research advanced, scientists have had to modify the definition of "prion." They learned that prions can be noninfectious as well as lethal, existing normally in the body. The protein has a Jekyll-and-Hyde personality: a normal form, required for healthy cellular functions, and a misfolded, pathological shape, which can kill the cell.

Like bacteria, viruses, and other conventional disease agents, the pathological prion kills by making more of itself and overwhelming the body's defenses. But the prion replicates in a different way, by converting the normal prion protein. These newly transformed prions "inherit" the original prion's pathological properties and ability to recruit others. Its *modus operandi* has evoked comparisons to Kurt Vonnegut's ice-nine, the permanently solid version of water that, on contact, progressively freezes liquid water into the same unmeltable entity.

The idea that a protein could pass inheritance features violated the orthodoxy of molecular biology. Anything that reproduces, such as bac-

teria, viruses, yeast, and healthy human cells, needs nucleic acids —
DNA and RNA — to do so, but prions go against this rule. Because they
don't have nucleic acids, they are impervious to the kind of assaults that
rip apart the genetic material of microbes. In fact, disease-causing
prions have been called the perfect pathogen. Their near-indestructi-
bility is one factor. Another is their quietly sinister ways. An individual
with a prion disease can live symptom-free for years, even decades.
There's no fever, no coughing, no elevated white-cell counts — no sign
whatsoever of an infection. But the prion is at work, converting normal
prion proteins as it ice-nines its way through the brain. Eventually, the
disease asserts itself. The memory starts to go. Senses like smell and
sight may vanish. Clumsiness and muscle twitching develop. For some,
permanent insomnia sets in. Once the disease is unleashed, the progno-
sis is set. There is no known treatment. Death inevitably ensues, within
a matter of months in some cases, more than a year in others. Because
prion diseases typically leave holes in the brain so that it resembles a
sponge, they are also called spongiform encephalopathies. Because they
can spread to other individuals, they are classified as transmissible.
Transmissible spongiform encephalopathies, or TSEs.

Like many others, I was only dimly aware of transmissible spongiform
encephalopathies in the 1980s. After all, TSEs were rare, occurring in
about one in a million people in the form of Creutzfeldt-Jakob disease
(CJD). Or it was a disease of historical and cultural interest called kuru,
which afflicted cannibalistic tribes in New Guinea in the 1950s. Or it
was of concern to the wool industry when sheep came down with a TSE
called scrapie.

But shortly after I joined *Scientific American* in 1989, Britain's mad
cow disease crisis and its attendant public health fears reached a
fevered pitch. Livestock feed made from the carcasses of sickened
sheep and cattle spread the mysterious illness like wildfire, creating
droves of "mad cows" that staggered about and lunged aggressively at
people. By early 1993, 1000 cattle a week were coming down with the
illness, officially called bovine spongiform encephalopathy, or BSE.
Many thousands more were being slaughtered preemptively and incinerated in an effort to control the nightmare that caused Britons to

recoil from beef—every sausage seemed sinister, every meat pie suspicious. Other nations began to seal their borders against British beef. But these trade bans did not keep the disease from spreading beyond the island to most of Europe and to Japan.

During the BSE crisis, British authorities tried to reassure the public, claiming that the disease could not infect people. Scientists assumed that because humans have been living with scrapie-infected sheep for more than two centuries, BSE didn't pose a threat to human health. But after young people began dying from a new strain of CJD in 1995, government officials realized they had made a terrible mistake.

Today, epidemiologists are still trying to determine how many people will ultimately come down with the new strain called variant Creutzfeldt-Jakob disease (vCJD). It could be a few dozen, several hundred, or a few thousand, or even tens of thousands. Moreover, the risk isn't restricted to British citizens—travelers who spent a few months in the country between the early 1980s and the mid-1990s are considered at risk. So are the approximately 5 million American military personnel who served in Europe between 1980 and 1996.

Often, upon about learning of BSE, people will remark that they are glad they like their beef well-done. But, unlike *Salmonella* or *E. coli*, the mad-cow prions cannot be cooked out. To inactivate the prions, you would have to incinerate your burger down to ash or soak it in plumber's lye—neither of which is a palatable option.

Just as it is wrong to think that a well-cooked burger is a safe burger, it would be a mistake for Americans to consider BSE an isolated foreign problem. True, the European Union and Japan have borne the brunt of mad cow disease because they imported infected cattle feed even during the BSE crisis. But BSE has spread far beyond those regions. Many countries in eastern Europe, northern Africa, the Middle East, and southeast Asia would probably find BSE if they had proper surveillance systems.

Although no BSE cases have been uncovered in the U.S., officials say the risk cannot be ignored. The General Accounting Office, the investigative arm of Congress, reported in January 2002 that significant gaps in protection exist that could allow the illness into the country. Should it happen, the economic cost to the $56-billion-a-year beef industry would be staggering. If the crisis got as bad as in the U.K., it would cost $15 billion in lost revenue alone. Japan estimated that its first three cases of mad cow disease cost $2.76 billion. What's more, it's not beef alone that has prion researchers worried. Cow material has been used in vaccines, dietary supplements, and other products not normally thought of as bovine-related.

More relevant to the U.S. than mad cows, perhaps, are the mad deer and elk loose in some parts of the country. Actually, they do not really go "mad"—they waste away, turning into fur and bones before they die. The affliction is called chronic wasting disease (CWD) and is the only known prion disease that affects wild animals. And because it is easily transmitted, it poses the potential for a rapid, uncontrolled spread. The CWD outbreak in Wisconsin seems to be particularly severe; left unchecked, the disease could threaten all the white-tailed deer in eastern North America. To avoid this end, Wisconsin authorities ordered a massive hunt in 2002 in an attempt to wipe out 25,000 deer in a 411-square-mile "hot zone" to prevent CWD's spread. But the culling may have come too late: In January 2003, a few CWD-infected deer were confirmed to have lived outside the zone.

Whether humans can get a prion disease from CWD venison is not known, although theoretically it's possible. Test-tube studies have shown that CWD prions can convert the normal human prion protein into the pathological form about as well as BSE prions do. The Centers for Disease Control and Prevention investigated the deaths of several Creutzfeldt-Jakob disease patients who dined on venison, but the agency did not find any definitive connection. Still, the CDC recommends avoiding the meat from an infected deer or elk.

Considering their rarity, prion diseases might have remained in neurobiology's backwater had it not been for the epidemics of mad cow and chronic wasting disease. Interestingly, these prion diseases have become problems because of human actions. BSE would never have emerged had cows not been fed slaughterhouse remains in order to speed their growth and boost their output of milk. Chronic wasting disease might have remained confined to a patch of land in the Rocky Mountains had it not been for the interstate trade of deer and elk for game farms.

One of the goals I have in mind in writing this book is to present the information on TSEs and prions so that you, the reader, can decide on the level of risk you are willing to accept.

As for my own personal risk assessment—whether to eat beef as I traveled around the U.K. during my book research—I decided to ask scientists what they did. Who better to question, I thought, than those with the most knowledge about the disease? Before I left, I had lunch with three U.S. scientists and asked for their thoughts. Sure, they responded, no problem at all; they would not hesitate eating British beef today. The U.K. sees only a few BSE cases a week on average, and the country has instituted several regulations that have enhanced food safety.

Still, I decided I would rather avoid U.K. beef. After all, I would only be there for a couple of weeks—why bother taking a minuscule risk when I could take a zero risk instead? I could hold off any cravings for a medium-rare porterhouse until I got back to New York.

My strategy hit a wall on an early-morning flight from Edinburgh to London. I had to dash out of the hotel before breakfast was served, and being in a somewhat remote neighborhood, I didn't have any choices along the way. Besides, I thought, there was always airplane food. Sure enough, after we achieved cruising altitude, a typical English breakfast arrived, complete with eggs, a slice of tomato—and a link of sausage. I considered leaving the meat on the tray. But then I thought about what the three scientists had said, and hunger got the better of me. So down it went.

Over the next few months, as I met with other U.S. researchers, I posed the same would-you-eat-beef-in-Europe question. Sure, I'd eat a sirloin steak in England, one scientist remarked. But then he added that he would avoid the T-bones and "mystery meats" that go into pasties, burgers—and sausages. Another researcher confessed to staying away from certain European meats but asked me not to reveal the choice for fear of some political implications. The scientist was especially wary of mechanically recovered meat, like that found in sausages. Another scientist gave me a similar opinion: yes to muscle meat like steak, no to ground-up meat like sausage.

Great—*now* you tell me. Still, my chance of catching mad cow disease after eating that piece of processed meat is undoubtedly negligible, if even that much. I face much greater health risks in my everyday life, considering my habitual jaywalking and the bus fumes I inhale on Madison Avenue. And I would assuredly be worse off if I failed to go the dentist regularly.

Yes, the risks are low—for now. As a man-made disaster, BSE proved that human activity can unpredictably alter the characteristics of prions so that they become dangerous to people. The prions behind chronic wasting disease appear to be following a similar trajectory as the illness spreads among wild deer. What will happen as those prions encounter other species is a mystery—and a threat.

CHAPTER 1

A Death in Devizes

An unusual death in the U.K. marks the arrival of a harrowing new brain disease.

A new resident arrived at Dunstan House Nursing Home in early May, and he was clearly an anomaly. Of course, like many of the elderly in this place, Stephen Churchill was here because he could no longer care for himself. Round-the-clock nursing was required to feed him, bathe him, and take him to the bathroom. Dementia had set in—he could not remember events just moments after they occurred. He had become largely unresponsive, and he spent most of his days lying in bed. At times, he could be seen in the sitting room, slumped in his wheelchair—he could no longer walk on his own, because his unsteady gait would send him careening.

Ordinarily, Stephen Churchill would seem to be simply another sad resident making a last stop in an old-age home. But he was no frail senior ravaged by time. Stephen Churchill was only 19 years old.

Stephen died two-and-a-half weeks after he had arrived, on Sunday night, May 21, 1995. The immediate cause of death was bronchopneumonia, fairly common for victims of neurological disorders. But the real illness probably began some ten years before, or perhaps even longer, and slowly ate away the healthy teenager's brain. The disease would go on to kill dozens more and create a worldwide shudder that many of the 60 million citizens of the U.K.—and anyone who visited there—might suffer Stephen Churchill's fate.

A Boundless Future

Stephen's end as a frail, mentally incapacitated young man seemed almost inconceivable for a boy with such a promising future, as his parents told me on a cool October morning in the living room of their three-story house in Devizes, a rural town in Wiltshire county about 90 miles west of London.[1] The Churchills' home, which doubles as a bed-and-breakfast, stands between the surprisingly busy, two-lane Bath Road and the Kennet and Avon Canal. They moved into this red-roofed white house, fronted by a knee-high brick wall, a few months before Stephen died.

Stephen was one of those bright children who became disruptive when bored with unchallenging class work. "We decided Stephen needed to be moved or he was not going to achieve at school," recalled his mother, Dorothy "Dot" Churchill, of their days in Stockton-on-Tees, near Middlesbrough in the northeast of England. Transferred to Red House, an independent school that prided itself on strictness, the seven-year-old Stephen thrived. He displayed an aptitude for languages, studying French and Latin. Later on, he would serve as the family's translator on trips to France.

Stephen's father, Dave, was promoted to the Wiltshire Fire Brigade in 1988, so he, Dot, Stephen, and daughter Helen moved some 300 miles south to Devizes. "We were pretty ordinary, Mr. and Mrs. Average Britain with two children and a dog," Dot said. Stephen ran into the same kind of trouble at the local comprehensive school, so the Churchills sent him some 20 miles west, to the private and more challenging King Edward's School in Bath.

Like many teenage boys, fighter jets captivated Stephen, and becoming part of the Royal Air Force would be his dream. In preparation, at age 13 he joined the local squadron of the Air Training Corps—a kind of a winged version of the Boy Scouts of America, but funded by the armed services. As cadets, members learned about aircraft and how to move about in camouflage. Stephen quickly rose up the ranks, achieving both the status of flight sergeant and the responsibility of training new recruits. In December 1993, at age 17, he applied for a scholarship from the Royal Air Force. The application wasn't simply filling out papers and collecting some recommendation letters. "He went and had four days of air-crew testing," Dave recalled. "Psychological, physical,

Stephen Churchill, April 1994. (*Photograph courtesy of David and Dorothy Churchill; Crown copyright material is reproduced with the permission of the Controller of HMSO and the Queen's Printer for Scotland.*)

coordination, leadership—you know, you've got a pile of barrels and some planks and you got to get it from here to there in swamps."

The testing revealed he had all the right qualities: the intelligence, the ability, the reactions, the background, the mental stability. "He came out super for air crew. He could be a pilot or navigator for fast jets. We're talking about the *crème de la crème* now. He had all the boxes ticked for air crew, pilot, or navigator," Dave proudly stated. Unfortunately, most of Stephen's 5-foot, 11-inch body was concentrated in his legs—in fact, he set the record there for longest thighs, according to Dave. "The training aircraft the RAF used, the Hawk, has a particularly tight cockpit. And if you pull the lever and eject, if your legs are longer than a certain length, you'll actually take your knees off." Stephen's legs were about an inch too long, preventing him from piloting.

Nevertheless, the Royal Air Force was still eager to have Stephen join and invited him to experience the life of a fighter controller, a high-stress job that involves coordinating the jets in the skies from sentry aircraft. Around Easter of 1994, the RAF sent him on a two-week training session to learn air navigation. During that time, on April 14, Stephen turned 18, which is the legal drinking age in England. That undoubtedly made it easier for Stephen to indulge in one of his hobbies—collecting unusual beer bottles from all over the world, which he stored in a box in his bedroom.

Troubling Signs

The problem child was growing into a focused young man. Nothing was going to stop him—at least nothing that Dave and Dot could see at the time. In hindsight, they remembered a few things out of the ordinary. Dave's father, an RAF navigator himself during World War II, died of cancer on May 1, 1994, at age 79. "Steve didn't show an awful lot of emotion," Dot said, even though he was close to his grandfather. In June, Stephen did miserably on his end-of-term school exams. On August 5, his parents sent Stephen to Salamanca, Spain, for a two-week language course, where he evidently enjoyed himself a bit too much. Over the course of his trip, Stephen fell off a loudspeaker at a disco and had a case of food poisoning from a bad chicken burger. "So when we met him in London, he was looking a bit thin," Dot recalled. The family attributed it to food poisoning and fatigue from travel.

Back in Devizes, Dot granted Stephen and a friend permission to borrow her Ford Fiesta for an evening out on August 26. An hour later, the Churchill's telephone rang. It was the police. Just a few miles away, Stephen had crashed the little Ford. "He turned a bend and came head-on to an army lorry coming the other way," Dot said of thecollision that hit the truck's driver's side. "He could never explain why he ended up in the middle of the road. How he wasn't killed God only knows." Long legs and a seat set back from the steering wheel helped, Dave reasoned.

The Ford was a total loss. "I was a bit peed off," Dot said, because it was her first good car and because she needed it for her job—running a charity that required visiting people with physical disabilities throughout the county. "And of course we couldn't get insurance for him to drive, so he couldn't drive anymore. He had to get his push-bike out. So it was all a very different atmosphere to what he'd been used to, nipping out in my car and taking the girls out and things. It wasn't going to happen anymore."

Life for Stephen soon soured even more. He did poorly on his exams—including, strangely, French. In an exam that tested vocabulary, grammar, and comprehension, he scored only 13 percent when in fact he was fluent in the language. "It was confusing to us and devastating to him," Dave said. "We would expect him to get into the high 70s, well into the 80s, and he'd be disappointed if he hadn't got a 90."

Even happy occasions did little to lift Stephen's mood. The Churchills celebrated their silver wedding anniversary with a large party on October 4, 1994, and Stephen organized the fireworks, which he obtained from the hardware store where he worked part-time. Stephen remained morose throughout the evening, and he staggered and even seemed a bit drunk when he was setting off the display.

Later that month, teachers at the King Edward's School reported to his parents that Stephen was not coping very well—he was not talking to his friends and had become isolated. At home, he often elected to stay in his bedroom rather than to go out as he usually did. He began losing weight, too. Depression among teenagers isn't uncommon, and a general practitioner in Devizes who examined Stephen told the family not to worry. Pull up your socks and knuckle down and do some work, she told Stephen. But it was clear that motivational words weren't going to do the trick. His attitude did not improve. "His shoes were dirty, he wasn't brushing his teeth—just wasn't coping on an everyday basis," Dot recalled. On October 18, Dave and Dot found a note in Stephen's bedroom addressed to them from him; it expressed sorrow for his not being able to handle life. Twice that month, the school had summoned the Churchills to discuss Stephen's academics, which were falling well behind expectations.

"Then, in the middle of November, I got a phone call from the headmaster," Dot said. The school was wondering where Dot was—after all, according to Stephen, she had agreed to allow Stephen to leave school, and she needed to sign the papers. "I think you've got the wrong boy here. You're talking about someone else," Dot told him. No, the headmaster replied, it was Stephen.

Dave got out of work, and the Churchills tore off to Bath. School officials found Stephen sitting in the school library, apparently in a daze. "I'm sorry, I'm sorry," he kept saying.

Stephen had reasons to be depressed—the death of his grandfather, poor grades in school, the car accident. "At that age in the British schooling, there's peer group pressure, competitiveness. It's easy to step back and say, well, that's depression," offered Dave, who took Stephen out to a local pub to talk that night. If you don't want to go to school, that's fine, Dave had told him—sweeping the streets would be better than being under academic pressure if it meant his health and well-being. After a couple of half-pints of beer, Stephen seemed to have become drunk. "He was used to alcohol," Dot said. "But here he was, after two half-pints of lager, legless."

The next morning, November 19, Stephen woke up bright and early and told his parents he was going into town to get a job. He had held a Saturday job in a Devizes hardware shop since he was 16, and he said he hoped to get a full-time position there. Two hours later, he returned and indicated that he had landed an interview at the hardware store. "A bit later on," Dot recalled, "he said, 'I've got an interview with the jewelers in town'." He told them where this jewelers was, much to Dave and Dot's bewilderment. "He described a place that doesn't exist," Dave said—it was just a parking lot.

Stephen continued reciting the events of that day, at least how he remembered them. He had gone to police headquarters and was going to get a job flying a helicopter. And he was to start tending bar at a pub down the road. "By then, things were becoming a bit stupid," Dot said. "He was pulling known facts together to create a fantasy" filled with details, Dave added. "He described where this building was, and there was not a building there, never mind a jewelers. However we challenged him, he was adamant that it was there, and he'd been there, and he had got the job and was starting on Wednesday or whatever. The alarm bells were ringing."

Another general practitioner examined Stephen, and in part because Stephen mentioned the death of his grandfather, the doctor concluded that depression was the cause and prescribed antidepressants. Such drugs work by regulating the signals between neurons, compensating for abnormal amounts of chemicals such as serotonin that neurons use to communicate with one another. Unfortunately, the antidepressants turned out to have no effect.

Stephen continued to deteriorate. The young man who danced on loudspeakers and dreamed of flying RAF jets was content to plant himself in front of the television all day. He started identifying with the programs in unusual ways. Watching *Baywatch*'s Pamela Anderson swim underwater made him think he was drowning. Cartoons frightened him into gripping the armrests. Dot would take him out every day just to keep him from watching television all the time.

"Within a few days I decided I would go to a supermarket in Chippenham, a town further north of here. To keep him out, we went for coffee and a donut. As we came out, I said to him, 'Well, what did you think of what we just had to eat?' He realized he couldn't remember. He'd eaten a donut, and he had had a tea, and he couldn't remember. We put it down to the effects of the drugs he was on."

Soon, Stephen grew fearful of water and sharp objects, and he refused to wash or shave. He became very withdrawn and lost some 25 pounds off his originally lean 145-pound frame. His hand-eye coordination was shot—he couldn't put food in his mouth even when it was already on the fork. He would reach for a cup, miss it, but continue with the action of bringing it to his lips. Simple tasks, such as signing his name or unlocking a door, escaped him. "He could tell you the date of D-day, capitals of the world, but couldn't tell you what day it was or what he had for breakfast," Dave recalled. Occasionally he suffered from visual and auditory hallucinations.

With Stephen's continued decline, a physician advised the family to place him in a psychiatric hospital. But with Christmas fast approaching, Dave and Dot decided he should remain at home at least until after New Year's Day. "Christmas was a big thing to Steve, and Helen was coming home from the university for the holidays, anyway," Dot recollected. "There was a set formula: a real tree with lights on, a particular menu for Christmas. Out of all of us, he was the one who had to have all these things done. But in fact, he fell asleep halfway through Christmas dinner. He hardly ate anything. He was on another planet, really. He just didn't know what was happening." A pair of fuzzy gorilla slippers— Helen's gift to Stephen—frightened him. It must be the drugs, the family rationalized.

"You Don't Die of Depression"

By January 3, 1995, the Churchills decided that the local psychiatric hospital was the best place for Stephen, who by now weighed 120 pounds and would eventually lose 30 more. The staff could monitor him and figure out exactly what was wrong and what combination of drugs might work. He was given antipsychotics such as sulpiride, a compound highly effective in controlling symptoms of schizophrenia, such as hallucinations and lethargy, but they failed to help. Stephen had become very compliant—if you told him to stand by the front door, he would do it without question. "Here was a guy who would have argued that black was white 12 months previously," Dave said. "It was a sea change in attitude." His gait deteriorated so much that two assistants had to help him walk; eventually, he was moved about in a wheelchair. Occasionally, he would twitch and jerk his limbs—a condition that physicians refer to as myoclonus. He speech began to slur (a condition called dysarthria), and the nursing staff couldn't follow his train of thought.

After three-and-a-half weeks in the psychiatric hospital—including a medical emergency that may have been caused by an overuse of drugs intended to elicit a response, which put Stephen at death's door one morning—it became clear to the hospital physicians that Stephen wasn't suffering from a psychiatric problem but a neurological one. At the end of January 1995, Stephen was transferred to the Royal United Hospital in Bath, under the care of neurologist David Bateman, and over the course of several days, a battery of tests began. They included an electroencephalogram (EEG) to measure the electrical activity of the brain, a "spinal tap," in which fluid from the base of the spine is withdrawn and examined, and blood tests.

Some of the tests revealed abnormalities: the nerve response in his leg muscles was off, and the EEG showed unusual slow-wave activity, but nothing that would indicate a particular disease. The cerebrospinal fluid revealed no sign of infection—it ran clear, as it would for a healthy individual. Blood tests presented nothing remarkable, and Stephen didn't have a fever. There were simply no signs of infection.[2]

"We weren't told anything until February 13. The night before, Dr. Bateman said he'd like to see us in his office," Dot remembered. Finally, they would have some answers, the Churchills thought. They weren't prepared for the news. Stephen, the neurologist told them, had a progressive degenerative neurological illness and couldn't be saved. "We just sat there, because we hadn't thought it was terminal. We were still thinking it was depression here, and you don't die of depression," Dot said. "You can't put it into words how you feel."

No Answers

A few days later the entire family went via ambulance to the National Hospital of Neurology and Neurosurgery in London for a second opinion. All they brought were Stephen's brain scans—no notes or test results. Stephen began a new barrage of tests, including those for HIV and Huntington's disease—"You name it, he had it," Dot said. "They even looked into whether he'd been in contact with chemicals, because he'd been at Salisbury Plain, the site of the air-training corps." They looked into possible infections from his trips abroad, and Dave and Dot had to construct family trees to check whether there might be something in the family's genetics "which caused a family rift," Dot explained.

With no firm diagnosis, physicians conducted a brain biopsy—a last-ditch procedure wherein a small hole is drilled in the forehead and a microscopic bit of the right frontal lobe is removed for examination, at the risk of causing the brain to hemorrhage or an abscess to form.[3] It could also cause changes in personality. The hope was that it might reveal the presence of a tumor or inflammation of blood vessels in the brain—which would be good news, since those conditions might be treatable. Under the microscope, unfortunately, the sample revealed little.

Deeper inside, however, where tests and biopsies couldn't reach, neural tissue was crumbling, leaving behind microscopic holes that made Stephen's brain look like a sponge. This spongiform change did not extend much into Stephen's cerebral cortex, the 1/8-inch thick, gray outer layer of the brain that controls higher mental functions such as language and conscious thought. That the cerebral cortex was mostly intact suggested that Stephen's speech impairment was rooted in muscle coordination, not in damage to his brain's language center. He knew what to say but couldn't coordinate his larynx, lips, and tongue to form the words.

Three weeks later, the Churchills were no closer to an answer. "We had to demand an interview with the neurologist," Dot recalled. "She didn't want to see us, really—you could see that. She said they didn't have any results really back and didn't know what was going on. But in her opinion, Stephen had years in single figures to live. So of course as parents we thought nine years."

Stephen was transferred back to the hospital in Bath. Because of insufficient staffing, the Churchills took it upon themselves to provide for Stephen's care between 8 A.M. and 8 P.M. (they had found out that one day there wasn't staff available to feed Stephen, who by now had to be spoon-fed). Although Stephen responded to his parents and was happy to visit the family home once a week, he didn't seem to be aware of his growing infirmity.

Hospital officials told the Churchills that their son was probably too much of a handful to care for at home, so Dave and Dot began searching for a place for Stephen to live out his final days. By this time, the neurologist had told them that Stephen had about two years to live. "So in a fortnight, we had gone from nine years to two years," Dot said. There were no obvious places for young patients suffering from dementia. The Churchills spent weeks scouring the region for a suitable home. Many had waiting lists longer than Stephen's anticipated lifetime;

others, "I wouldn't put a pig in, never mind a relative," Dave said. Some were 60 miles away, too far for Stephen's friends to visit.

Several calls and follow-ups later, Dot and Dave came across John and Margaret O'Dea, who ran the Dunstan House Nursing Home in Calne, a small town only a few miles north. "It was filled with elderly people, but they had this one room that was slightly separate," Dot recalled. It was also next to a bathroom well-equipped to handle disabled patients. The plan was that Stephen would be in this nursing home at night, while Dave and Dot cared for him at home in Devizes during the day. "It took quite a while for this all to be set up. You've got to talk about finances, who would pay for all of this. In the end, we moved Stephen to this nursing home, near the fourth [of May 1995], thinking we've got a couple of years here," Dot said.

To make things seem more like home, the Churchills brought some of Stephen's personal belongings, including his beer bottle collection and his poster of a reclining Pamela Anderson in her red *Baywatch* swimsuit. The nursing home created a welcome atmosphere; even the residents talked to Stephen in the sitting room and treated him as a grandchild, although Stephen hardly replied. "At times we still feel he was aware right up until the end, because of certain things he responded to. He wasn't in a coma or a persistent vegetative state," Dot explained.

Stephen's stay was much shorter than anyone anticipated. Less than three weeks later, on May 21, 1995, Stephen died of bronchopneumonia. Respiratory ailments are a common cause of death in dementia patients. They often lack the proper muscle control for good cough and gag reflexes, so that the contents of the stomach can wash into the lungs, damaging delicate tissues and paving the way for infections from germs descending from the mouth and nose.

To discover what led to Stephen's neurological decline, the Churchills granted permission for a postmortem examination. No one had given them a specific answer, other than that he had suffered from a progressive neurodegenerative disease. The family's only clue had been gleaned at the National Hospital in London, when Stephen was recuperating from his brain biopsy. "When someone is recovering from an operation, you've got nothing to do," Dave explained. "You can't talk to him because he's still zonked. So you read everything in sight, including the notes. By then, we were very good at reading hospital notes. They left a full set of notes there, and on the outer cover of the file, it said the patient's name, age, date of birth, procedure." Then they saw three initials they had never seen before or had mentioned to them. On the entry to explain Stephen's operation, Dave said, the form listed at the top corner "the reason: C-J-D, question mark."

CJD — Creutzfeldt-Jakob disease. An illness of people in their sixties.

CHAPTER 2

One in a Million

A rare disease only gradually becomes recognized as the most common human spongiform encephalopathy.

Dr. Hans Gerhard Creutzfeldt (1885–1964) has a disease named after him, even though he probably never saw a patient with it. Born in 1885 in Harburg, Germany, a village later incorporated into Hamburg, Creutzfeldt received his doctorate from the University of Kiel in 1909. He apprenticed with various prominent researchers, including a stint with Alois Alzheimer between 1912 and 1914. Creutzfeldt held high-level posts in several neurology departments and defied the Nazi party by allowing his hospital to be used as "refuge for persons who had fallen foul of the 'hereditary laws'."[1] Today, he is honored by having a clinic in Kiel named after him. But he is best known for a 1920 article he wrote for a German journal, in which the then 35-year-old neurologist recounted the sad life of Bertha Elschker.[2]

Born in Silesia in 1890, Bertha, whose mother died in 1904, ended up in a Catholic orphanage. There, Creutzfeldt wrote, the young girl "stood out because of her childish and stubborn nature; she was lively and much occupied with dolls and childish play. She was [also] industrious at [school] work."[3] In the summer of 1912, Bertha was hospitalized for what doctors at the dermatological clinic of Breslau University described as massive scaling of the skin on her face, hands, abdomen, and feet caused by hysteria. Occasionally, her legs would jerk spasmatically, making walking difficult at times.

Her condition followed a remitting course, improving and then worsening. By May 1913, her unsteady gait had returned more severely than ever; she even fell over while standing. More seriously, her mental state changed suddenly. "She no longer wanted to eat or bathe . . . she assumed peculiar positions, in that she bent over to her left and pressed her hand against her heart," Creutzfeldt wrote. Bertha's menstrual periods became irregular, and she often spotted between periods. On June 17, "She suddenly screamed out that her sister was dead, that she was to blame, that she was possessed of the devil, that she herself was dead, that she wanted to sacrifice herself. . . . Only rarely were sensible answers to be obtained from her."[4] She became emaciated and could not walk or stand without help, and she mostly "presented a dazed, stupefied expression." She ran a remittent fever, and the muscles in her face and limbs jerked, tic-like. Soon epileptic seizures commenced, one right after the other. She was in a deep coma at the time of her death on August 11, 1913.[5] A postmortem revealed a host of problems wracking her body, including bronchopneumonia (probably the immediate cause of death), ovarian cysts, and congestive kidneys.

Bertha's condition might have escaped notice—World War I temporarily delayed Creutzfeldt's write-up of the case—had it not been for neurologist Alfons Maria Jakob (1884–1931). The son of a shopkeeper in Aschaffenburg, Bavaria, Jakob rose through the ranks of academia and, like Creutzfeldt, apprenticed in the labs of highly regarded medical men (including Alzheimer), eventually becoming head of the psychiatric state hospital of Hamburg-Friedrichsberg in 1914.

Over the years, Jakob would elucidate many ailments affecting the nervous system, including concussions, multiple sclerosis, yellow fever, and syphilis, and his reputation as a teacher drew students from Japan, Russia, Portugal, and the U.S. His contributions are all the more impressive because he accomplished them in a relatively short time—he died at age 47 after surgery had failed to save him from abdominal abscesses brought on by a bone infection.[6]

In 1921, Jakob was preparing his own paper on strange neurology cases when he came across a preprint of Creutzfeldt's paper. To Jakob, Bertha Elschker's condition bore striking resemblance to his own patients, to whom he referred with disguised versions of their real names: Hein, Jendross, Ernst Ka., Jac., A. Hoffert. Those patients, however, were older than Bertha, mostly in their late 30s and 40s.

> They display a lurid mixture of symptoms . . . so that on the one hand, they are reminiscent of multiple sclerosis and spastic disorders, and,

on the other, show certain relationships with the striatal disease processes such as pseudosclerosis [now called Wilson's disease, a hereditary condition in which copper builds up in the body and becomes toxic[7]] or the varieties of chorea. Associated with these symptoms are striking mental changes which complicate the disease picture even more.[8]

Based on his and Creutzfeldt's cases, Jakob also asserted that:

> I thought it possible to characterize the clinical picture of the disease as follows: this is a disease of middle and late life which begins with progressive—slowly at first—disturbances of the motor apparatus and of sensation. The patients complain of weakness and pains in their legs, which become stiff. In walking, their legs often give out under them and they fall down; meanwhile, objective findings on examination at first are lacking.[9]

But in postmortems, "on microscopic examination we find severe, extensive histologic changes throughout the entire central nervous system." The changes Jakob saw included loss of neurons, neuronal swelling, and the proliferation of astrocytes, those star-shaped cells that, in abundance, are usually a sign of the brain's attempt to repair damage.

Although Jakob called it spastic pseudosclerosis, the term Creutzfeldt-Jakob disease (CJD) came from German psychiatrist and neurologist Walther Spielmeyer (1879–1935). Spielmeyer, who had made a reputation for himself by describing the nerve damage sustained by soldiers in World War I, named CJD in a 1922 monograph on the histopathology of the nervous system. Given the rarity of the condition and the variability of the clinical picture, others who attended such patients over the years proffered different nomenclature for CJD, including "cortico-pallido-spinal degeneration," "presenile dementia with cortical blindness," and, in a sacrifice of brevity for detail, "subacute vascular encephalopathy with mental disorder, focal disturbances, and myoclonus epilepsy."[10] One researcher regarded CJD as a "dumping ground for several rare cases of presenile dementia."[11] New names continued to pop up as late as 1960. The name Creutzfeldt-Jakob disease finally stuck in the 1970s, after the publication of English translations of the pair's original cases.

The Unlucky Few

CJD only slowly emerged as a disease unto itself thanks to rigorous clinical observations and study by physicians, who can mistake the symptoms of CJD for Alzheimer's disease, multiple sclerosis, or a stroke. Once clinicians and pathologists understood what to look for, CJD wasn't as rare as they thought. Today, most cases are "sporadic"—they happen for no known reason, and 80 percent of those afflicted are between the ages of 50 and 70.[12] The mean age of death is in the middle to late 60s.[13] Mysteriously, the incidence of CJD starts to drop in the 70s, and the decline does not appear to be the result of poor surveillance.

Sporadic CJD affects about one in a million people each year, translating to thousands of cases annually throughout the world. Compare that with the state of knowledge in 1965, when all the CJD cases *ever* known to have existed numbered 150. A one-in-a-million rate is still rare—you are far more likely to die from Alzheimer's disease (180 deaths per 1 million people in the U.S.[14]). But at an estimated 290 a year, CJD deaths are much more common than deaths from lightning strikes (about 80 per year in the U.S.).

Just why a certain unlucky few get this invariably fatal disease, which generally takes its victims from first symptoms to death in about five months, remains mysterious. Studies indicate that it follows no seasonal pattern or geographic clustering and hasn't shown any change in incidence over the years. CJD affects all races and ethnic groups.[15] It appears worldwide except for central Africa, an observation that probably reflects poor surveillance and the comparatively short life span there—in some African countries, life expectancy is in the mid-30s because of AIDS.

Researchers over the years tried to isolate risk factors. Some found physical injury, others reported surgery with sutures. Other findings were contradictory: One paper reported that having pet cats elevated the risk; another found that contact with animals *other* than cats did the trick.[16] Other factors, including diet and occupation, have never been confirmed as contributing causes.[17] As best as scientists can determine, sporadic CJD just happens—a particularly frustrating conclusion for both victims and their relatives.

What is known is that between 10 and 15 percent of all CJD patients contract the disease as a result of inheritance. If one of your parents had the disease, then you stand a 50-50 chance of getting it yourself. Familial CJD strikes earlier in age and has a more protracted symptom period than sporadic CJD. It seems that people with no family history of CJD but who develop the illness do so because something triggers a similar change in their body. But despite intensive searches, no one has found out what pulls the trigger in those one-in-a-million individuals.

Diagnosing CJD

Despite the accumulation of knowledge about CJD over the eight-plus decades since its discovery, making a firm diagnosis is still difficult. In the U.S., where only some states require it to be reported, there is no active surveillance program in the sense that a health-care professional visits the family of a suspected CJD victim and conducts extensive interviews. Rather, the Centers for Disease Control and Prevention in Atlanta collect mortality data from death certificates, checking the codes listed for CJD. This method finds 5,014 deaths due to CJD in the U.S. from 1979 through 1999, which makes for an annual rate on par with the worldwide estimate of one in a million for sporadic CJD.[18]

Although dementia, psychiatric and behavioral problems, muscle twitching (myoclonus), and incoordination (ataxia) are typical signs of CJD, not all patients display all of them. In rare cases, patients have seizures; others go blind because the visual centers in their brains are destroyed, not because their eyes malfunction. Lab tests reveal no inflammation or consistent abnormalities in the liver or urine. Blood tests present no signs of antibody production that would signal the body's attempt to combat an infectious organism. Some CJD patients in their later stages show a characteristic pattern of periodic spikes on their electroencephalograms (EEG), but that reflects general brain damage and by itself is not considered diagnostic. The cerebrospinal fluid is also largely normal, although sometimes there are elevated levels of the so-called 14-3-3 protein.

Actually a family of proteins, the 14-3-3 protein was first discovered in 1967 and given a numerical designation based on where it was ulti-mately found after repeated purification steps that separate a sample into different fractions. Specifically, it emerged as the 3rd fraction separated from the 3rd fraction taken from the 14th fraction isolated from a brain tissue sample. In the brain, 14-3-3 regulates many cell functions, including signaling between cells. But with brain damage, the protein may leak down the spinal column. The presence of 14-3-3 in the cerebrospinal fluid isn't specific to CJD; other types of damage to the central nervous system, such as stroke and herpes simplex encephalitis, also raise the levels.

The only surefire way to diagnose CJD is to look at brain tissue. Under the microscope, CJD patient brains have a very notable sign— the presence of holes, reminiscent of the texture of a sponge, along with neuronal loss and gliosis (the proliferation of the neuron's helper cells, such as astrocytes). Spongiform change especially has become the unmistakable marker, and for this reason, CJD is classified as a spongiform encephalopathy. To help make the holes stand out, researchers will add stain to the brain tissue, because cell structures are mostly colorless. Some of these stains are the same as those used to dye clothing. Common ones include Congo red, eosin, silver, and iodine. Derived from plants, minerals, and substances such as coal tar, the dyes show detail not unlike the way a wood stain reveals the grain of an oak tabletop. In the case of spongiform change, holes leap out once cells take up the dye.

Postmortems of CJD cases, however, are not *de rigueur* in the U.S. In fact, only about half of the suspected CJD cases are autopsied. Given sporadic CJD's one-in-a-million incidence and adding in familial forms, there should be a bit more than 300 CJD cases a year in the U.S. Yet the postmortem surveillance figures for 2001, compiled by Pierluigi Gambetti of Case Western Reserve University in Cleveland, reached only 154.

Using observable holes in brain tissue as the basis for CJD diagnosis, two noted researchers from the National Institutes of Health, Colin L. Masters and D. Carleton Gajdusek, reexamined the stored tissue samples from Jakob's patients. They stained them with eosin, which made the holes more apparent than the cresyl violet stain Jakob had used originally. Masters and Gajdusek concluded that only Jakob's third and fifth cases—those of Ernst Kahn (Ernst Ka.) and Auguste Hoffman (A. Hoffert)—suffered from CJD. (Jakob in fact did notice spongiform change in Hoffman's brain tissue but did not emphasize it.) Bertha's remitting symptoms and family history—she had two sisters who themselves were committed to mental institutions—argue against

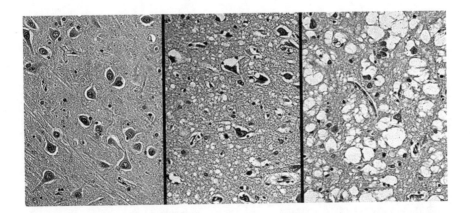

Microscopic holes puncture neural tissue in Creutzfeldt-Jakob disease, thereby giving parts of the brain a spongy look. The panels show increasing degrees of spongiform change, from none to severe (left to right). Large, tear-drop dark spots are the nuclei of glial cells, which support and protect neurons. Nuclei of neurons are visible as small dark spots. (*Herbert Budka, University of Vienna.*)

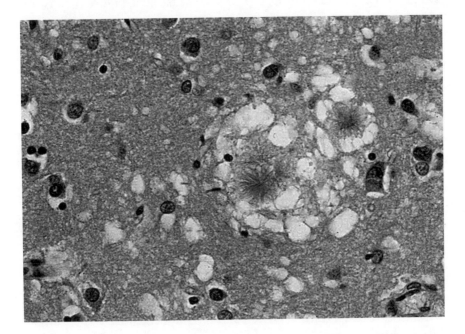

"Floral plaque" such as this one (the dark area surrounded by a ring of holes) appeared in Stephen Churchill's brain and was unlike anything seen before. (*Herbert Budka, University of Vienna.*)

CJD. "Creutzfeldt's case probably can be excluded from classification as a spongiform encephalopathy on the basis of his own clinical and pathological descriptions," Masters and Gajdusek concluded.[19] Bertha probably suffered from a demyelinating disease similar to multiple sclerosis, in which the fatty sheaths around nerves become frayed, leaving nerve fibers unprotected, or from a toxicity stemming from the ingestion of some sort of heavy metal, such as lead or mercury.[20]

Stephen's Case: CJD?

Stephen Churchill did indeed have the spongiform changes typical of CJD. Factor in the dementia and the muscle twitching, and a CJD diagnosis would seem to be assured. But there were discrepancies. Although his EEG was abnormal, it didn't match those of CJD patients. His early onset psychiatric illness—the depression, the anxiety, the delusions— were unusual for CJD. The course of his disease lasted more than twice as long as those of typical CJD patients. His age, too, was remarkable— at 19, he was some 40 years younger than the average patient. Only four other teenagers had ever been diagnosed with CJD.[21]

Stephen could have been chalked up as an odd case, the anomalies waved away. Yet there was something that couldn't be dismissed. Throughout his cerebellum were sticky protein clumps, called amyloid plaques. These plaques are rare in CJD, appearing in about 5 to 10 percent of the cases. *Amyloid* means starch-like—the plaques tint and react chemically in a manner similar to starch. Several types of stains can make these plaques visible—for example, eosin turns them a pale shade of pink. These protein clumps, though, have a special affinity for Congo red dye. They soak it up and appear as either green or gold, depending on the view through a polarizing filter.[22]

Several neurodegenerative conditions produce amyloid plaques. Alzheimer's disease leaves behind sticky clumps made of beta-amyloid and tau proteins. A protein called alpha-synuclein fills the brain of Parkinson's disease patients. Even an aged but otherwise healthy brain may harbor a few amyloid plaques.

These plaques are tough. Researchers typically dissolve such protein masses with biological compounds called enzymes. Living things produce all sorts of enzymes to assist necessary biochemical reactions—creating pepsin and trypsin to help break down food, for example. To dissolve proteins in the laboratory, researchers rely on enzymes called proteases—the -*ase* denotes its enzymatic activity. There are all sorts of proteases that chop up proteins in different ways and to different degrees, but a favorite is one called proteinase K. Derived from a fungus, proteinase K is one of the most broadly powerful enzymes. Yet it hardly does anything to the protein plaques of CJD.

But Stephen's plaques were different. They seemed to be surrounded by holes, like petals around a flower. These "floral plaques" had never before been seen in CJD patients. The characteristics of the plaques in Stephen's brain do bear a striking resemblance to another type of protein mass. That type was only seen half a world away, in aboriginal people living in the Highlands of Papua New Guinea—in the brains of cannibals.

CHAPTER 3

The Cannibals' Laughing Death

On a South Pacific island, two pioneering researchers begin to unlock the mysterious epidemic of kuru.

New Guinea practically teeters on Cape York, the northernmost tip of Australia. Hemmed in by the equator and the 10° line of south latitude, the island spans some 1500 miles in length and 400 miles in width. At 309,000 square miles, New Guinea covers as much area as the Atlantic states from Maine to South Carolina and is more than three times the size of Great Britain. Rugged mountains lining the island's center stretch nearly the entire length. Dense rain forests coat the surface and are occasionally cut by broad valleys and waterfalls, some of which drop hundreds of feet, and some so hidden they only reveal themselves by their thunderous roar and rising mist. Swamps filled with algae-covered mangroves and reeking black mud rim much of the coast.

Early on, sailors knew little of the inhabitants of New Guinea except to fear them. Spanish and Portuguese explorers of the sixteenth century were the first Europeans to set foot on the island, and they no doubt contributed to the myths surrounding the primitive societies there. New Guineans were thought to be fierce, relentless warriors adept with the bow and arrow, the stone axe, the spear.

But as contact increased, the belief that the natives were nothing more than violent savages melted away. Most encounters were peaceful. In fact, some New Guineans ran the other way upon seeing white men. In the 1930s, Europeans and Australians found nearly a million inhabitants of the mountain Highlands not previously known to the rest of the world. To the Highlanders, the pale visitors were dead relatives returning.

New Guinea has been home to an incredibly diverse assortment of people—hundreds of ethnic groups speaking upward of 700 languages now populate the land. Such variety was helped along by European colonization of the island that began in the nineteenth century. It started with the Dutch, who took control of the westernmost portion of New Guinea. By 1884, Germany controlled the northwest part, while Britain, and then Australia in 1906, administered the southeastern portion. Today, the western half of the island belongs to Indonesia, as the province of Irian Jaya; the eastern half achieved independence as Papua New Guinea in 1975.

That some of the natives were cannibals became widely known once Europeans and Australians penetrated the interior. Cannibalism produces a visceral disgust in most people today, but there's plenty of archaeological evidence to indicate that our ancestors around the world engaged in some form of human feasting.

Rather than eating their enemies, as some cannibalistic tribes did, the 14,000 members of the Fore tribe (pronounced *for*-ay) of the Eastern Highlands dined closer to home: they ate their relatives. Following up on the work of Australian anthropologists Catherine H. and Ronald M. Berndt, Shirley Lindenbaum, along with her husband at the time, Robert Glasse, spent the early 1960s documenting the life of the Fore and the history of their cannibalistic practices. Lindenbaum, now with the City University of New York Graduate Center, described how the South Fore prepared the body for consumption:

> In the deceased's old sugarcane garden, maternal kin dismembered the corpse with a bamboo knife and stone axe. They first removed hands and feet, then cut open the arms and legs to strip out the muscles. Opening the chest and belly, they avoided rupturing the gall bladder, whose bitter contents would ruin the meat. After severing the head, they fractured the skull to remove the brain. Meat, viscera, and brain were all eaten.[1]

With their bare hands, the South Fore would squeeze the brain into a pulp and stuff the mash into bamboo cylinders for steaming.[2] "Marrow was sucked from cracked bones, and sometimes the pulverized bones themselves were cooked and eaten with green vegetables."[3] Even the feces in the intestines were consumed, after first being cooked with greens. The South Fore's cousins to the north would bury the corpse for several days, allowing the flesh to "ripen." That way, "the maggots could be cooked as a separate delicacy," Lindenbaum explained.

Not all bodies were suitable for consumption. The Fore avoided those who died of dysentery or leprosy, for instance. As to why they ate their own, Lindenbaum said, "There was no thought of acquiring the power or personality of the deceased. Nor is it correct to speak of ritual cannibalism. While the finger and jaw bones of some relatives were retained for supernatural communication, Fore attitudes toward the bodies they consumed revolved around their fertilizing, rather than their moral, effect." Because the dead who were buried encouraged bloom in the gardens, the Fore reasoned that bodies must have some sort of regenerative power. "The flesh of the deceased was thought particularly suitable for invalids," Lindenbaum explained.

Fore cannibalism probably began no earlier than the first quarter of the twentieth century. Glasse and Lindenbaum reported that the Fore insisted the practice started within living memory, and through extensive interviews concluded that it first began among the North Fore and then a decade or two later among the South Fore. Likely reasons for the onset of this custom included an increased population, overhunting, and the conversion of forests to farmland, circumstances that reduced the availability of wild game. The Fore subsisted primarily on cultivated yams, taro root, corn, and sweet potatoes; they supplemented their protein-poor diets with pigs they raised, and opossums, lizards, and other small game animal they occasionally caught. When meat was obtained, the men got the best parts and left the entrails for the women and children. Believing that women weakened them, the men lived separately and rarely shared their meat. For protein, the women and children ate grubs and insects and generally relied on human flesh as an important dietary addition. The men would occasionally consume the human flesh as well.

Epidemic in the Bush

Several reasons inspired Europeans and Australians to make their way among the Fore. Gold hunters followed streams of glitter into the mountains. Miners found mineral resources to be exploited. Missionaries saw a chance to civilize "savages." Anthropologists found new ways of life to document. Others came because they could indulge in behaviors deemed unsavory and illegal in the West: Fore were quite sexual and might greet one another by fondling the genitals. Children were sometimes sent to stimulate the elderly.

Establishing a colonial foothold in New Guinea meant caring for the inhabitants, who provided a pool of workers for the road building, mining, and other labor-intensive development projects. But the remoteness of the island made it difficult to attract and keep physicians. Australian officials were happy to get Vincent Zigas. Born in Lithuania, Zigas studied medicine in Germany and arrived in New Guinea in 1950 to treat rampant venereal disease. By 1955, Zigas was granted Australian citizenship, enabling him to work for the Department of Public Health and to remain in New Guinea, which pleased him. "Not all the natives are gentle, not all have discovered the arts, not all have grace and beauty. They are, as I discovered, also the plainest, most down-to-earth, average persons," he wrote in his posthumously published memoir.[4]

To Zigas, the company of Highland tribespeople was preferable to that of the Australian colonials with whom he occasionally hobnobbed at special gatherings. In the spring of 1955, out of a sense of obligation, he had come to one such gathering—a christening. He had, after all, helped deliver the child. Among the freely flowing liquor and the arguing of the genteel set, Zigas met John MacArthur, a patrol officer assigned to set up an administrative outpost in the North Fore part of the Highlands. Like Zigas, he had developed a deep respect for the Highlanders and resented others at the party who insisted that the Fore were lazy, libidinous, and irresponsible. MacArthur told Zigas of the health problems plaguing the Fore: the dysentery, the pneumonia, and the open sores of yaws (a bacterial infection). He also told Zigas of his encounter with a small girl shivering violently by a fire, jerking her head from side to side. She was a victim of sorcery, the villagers told him, and she would die in a matter of weeks. They called it *kuru*, meaning to tremble or shiver.

Three months later, MacArthur sent a guide named Apekono to take Zigas on the arduous trek from the hospital in Goroka to the Fore settlements further in the mountains.

> On our way to Hegeteru village, a two-day walk of six to eight hours each, we came across a small village, a mere handful of huts. We turned onto a faint track leading to a small, dilapidated hut that I had thought uninhabited. Apekono nudged me and pointed through the doorway of the hut. I stepped warily to look inside. On the ground in the far corner sat a woman of about thirty. She looked odd, not ill, rather emaciated, looking up with blank eyes with a mask-like expression. There was an occasional fine tremor of her head and trunk, as if she were shivering from cold, though the day was very warm.[5]

Thinking that her symptoms might be psychosomatic, Zigas tried his own kind of sorcery, waving a liniment tube over her and commanding her to get up and walk. "The woman struggled feebly as if to rise, then, exhausted, started to tremble more violently, making a sound of foolish laughter, akin to a titter."

It turned out that most of these cases occurred in the South Fore, and most victims were women. One quarter were children of both sexes. Kuru went through distinct and remarkably uniform phases. It typically began as a headache and pain in the limbs. Patients soon developed an unsteady gait and started to tremble with spasms. Then, patients could no longer control voluntary movements and lost the ability to stand up. Eventually, muscles for swallowing stopped working, so many victims began to waste away for want of food and water. Dysarthria—a slurring of speech—set in, probably resulting from damage to muscle timing areas of the brain.

Trapped in their unmoving bodies, many kuru patients developed bedsores, paving the way for other infections and gangrene. Even at the end, patients seem to remain cognizant, though mute and still. After the first symptoms appeared, death (which often resulted from pneumonia or opportunistic infections) followed in about 12 to 18 months in adults, and 3 to 12 months in children. Because muscle spasms made some kuru victims appear to grimace and chuckle, news reporters would soon refer to the disease as "laughing death." It provided titillating headlines, such as "Kuru—The Laugh That Brings Death," much to the chagrin of the doctors who studied it.

Kuru reached epidemic proportions and devastated the Fore society. Between 1957 and 1968, more than 1100 people died from kuru in a South Fore population of 8000, according to anthropologist Shirley Lindenbaum. At its height in the late 1950s, twice as many women died from kuru as did men; in some villages, the ratio was 3:1. The annual death rate was about 1 percent, or about 50 times the AIDS death rate in the U.S. in 1995, its peak year.

Motherless households became common; the practice of polygamy became rare. Men were left alone to care for children, and they had to raise the pigs and tend the crops as well as work their regular jobs in the mines and at construction sites. Virtually every adult male lost a loved one to kuru. Believing the cause was sorcery, many Fore attacked and killed those they suspected of witchcraft. Known as *tukabu*, the revenge called for strangulation of the sorcerer and using a stone to fracture the femur, shatter the ribs, and break the spine.

Of course, sorcery wasn't the cause. In the mid-1960s, a young Johns Hopkins University resident named Paul Brown was recruited to retrieve brain samples from kuru regions. At his office at the National Institutes of Health, the senior research M.D. recalled his time in New Guinea. "Everybody 'knew' that cannibalism was the cause of it," he remarked, referring to missionaries, bush pilots, and other outsiders. "It doesn't take a genius to realize that if you've got a disease reaching epidemic proportions in a group of people that are eating sick people, then a pretty plausible guess is that cannibalism is the cause of the disease," Brown said. "But we couldn't prove it" right away.[6]

A Real-Life *M*A*S*H* Doctor

Zigas made no headway in treating kuru. He tried sulfa drugs, antibiotics, vitamin B supplements, copper sulfate therapy—all to no avail. Patients continued on their inexorable decline. Tests on tissue samples sent to Australia came back negative for infectious agents.

By March 1957, Zigas wasn't sure of his next steps. All he knew was that he planned to return to the outpost at Okapa from his hospital base in Kainantu; he would be going with the new Australian administrator for the region, Jack Baker, and taking two female kuru patients from the hospital back to their homes.

The day before he was to leave, a curious stranger appeared at his door. Zigas described him in his memoir:

> At first glance he looked like a hippie, though shorn of beard and long hair, who had rebelled and run off to the Stone Age world. He wore much-worn shorts, an unbuttoned brownish-plaid shirt revealing a dirty T-shirt, and tattered sneakers. He was tall and lean, and one of those people whose age was difficult to guess, looking boyish with a soot-black crewcut unevenly trimmed, as if done by himself. He was just plain shabby. He was a well-built man with a remarkably shaped head, curiously piercing eyes, and ears that stood out from his head. It gave him the surprised, alert air of someone taking in all aspects of new subjects with thirst.[7]

It wasn't long before the peculiar man was looking at the two kuru patients to be taken back to Okapa, and machine-gunning Zigas with questions.

The visitor was 33-year-old D. Carleton Gajdusek (*Guy*-doo-shek), arguably the most colorful and eccentric character ever to win the medical Nobel Prize. Born in Yonkers, New York, Gajdusek was a child prodigy who stenciled in the names of prominent microbiologists on the steps to his attic chemistry laboratory. He was heavily influenced by his aunt, "Tante Irene," an entomologist who brought back artifacts and insect specimens from Asia. "Before I was ten years old I knew that I wanted to be a scientist like my aunt," he wrote.[8] He attended the University of Rochester and enrolled in Harvard Medical School at the age of 19. He spent much of the early 1950s traveling to South America, Iran, Turkey, Afghanistan, and various Pacific islands to study and treat rabies, plague, scurvy, and other "epidemiological problems in exotic and isolated populations," as he put it.[9] His many awards over the years, including his 1976 Nobel Prize, enabled him to bring to his U.S. home scores of youngsters from Melanesia. He adopted dozens legally and sent many more through school in New Guinea and the Caroline Islands.[10]

In 1954 Gajdusek went to work with Sir Frank Macfarlane Burnet (known as "Sir Mac" to his friends and colleagues) at the Walter and Eliza Hall Institute in Melbourne, Australia. In a handwritten letter dated April 1957, Burnet, a Nobelist himself, described Gajdusek's personality as "quite extraordinary and . . . almost legendary amongst my colleagues in the U.S. [Virologist John Franklin] Enders told me that Gajdusek was very bright but you never knew when he would leave off work for a week to study Hegel or a month to go off to work with the Hopi Indians."

Infectious diseases expert Joseph E. Smadel, then director of the Walter Reed Army Medical Center's division of communicable diseases, was Gajdusek's boss while Gajdusek fulfilled his military obligation at Walter Reed in 1952 and 1953. Smadel advised Sir Mac that the only way to handle him was a swift kick in the rear. Burnet concluded:

> My own summing up was that he had an intelligence quotient up in the 180s and the emotional immaturity of a 15-year-old. He is quite manically energetic when his enthusiasm is roused and can inspire enthusiasm in his technical assistants. He is completely self-centered, thick-skinned, and inconsiderate, but equally won't let danger, physical difficulty, or other people's feelings interfere in the least with what he wants to do. He apparently has no interest in women but an almost obsessional interest in children, none whatever in clothes and cleanliness; and he can live cheerfully in a slum or a grass hut.[11]

Having known and worked for Gajdusek for the better part of four decades, Paul Brown doesn't quite agree with Burnet's description. "No, Carleton is not a 15-year-old. A better way of saying it is that Carleton throughout his entire life almost succeeded in doing exactly what he wanted," Brown said. "He has such a verve and audacity and a lust for life, an intellectual life as well as a physical life. But he doesn't respond well to being told he can't do something. He simply doesn't. That is a childish quality, but he's pulled it off, he's got so much talent." Brown sums up Gajdusek this way: "Carleton is a real-life version of the doctors in *M*A*S*H*."

A Lifelong Pursuit Begins

Gajdusek was on his way back to the U.S. when he decided to stop at New Guinea in 1957. He was interested in the "study of child growth and development and disease patterns in primitive cultures."[12] Back then, before the era of world travel and globalization, researchers could learn quite a bit about the human immune system by studying pre-industrial island communities. Before Gajdusek allowed Brown to start his kuru fieldwork, Brown tested inhabitants of the Caroline Islands for antibodies to measles. The goal was to find isolated populations never exposed to measles, enabling scientists to determine if the measles vaccine would endure in the absence of an active, circulating measles virus. There would be no way to determine that anywhere else in the world, since most people have been exposed to measles or the vaccine.

Gajdusek was planning to do a few months of pediatric research when he landed at Port Moresby on the southernmost part of New Guinea, whereupon he learned about Zigas and kuru from Roy F. R. Scragg, the acting public health official. Zigas soon showed him the two female kuru patients at Kainantu. Gajdusek recalled:

> When I saw them, they were no longer ambulatory; and the tremors, athetoid [spastic] movement, and blurred speech all pointed to a chronic neurological disorder unassociated with any acute infectious disease at onset (or even in the months or years before onset) which was dramatic enough to be recalled by reliable informants in their community. They were rational, but articulation of speech was very poor. Silly smiles, with grimacing, were prominent. Fixed and pained facies and slow, clumsy, voluntary motion (apparently in an attempt to overcome tremors and athetoid movement) were prominent also.[13]

The next day, they drove four hours with the two women in the back of the jeep, arriving in Okapa in a soaking rain and setting up a rudimentary laboratory in the home of a patrol post officer.

On March 15, 1957, Gajdusek wrote to Joe Smadel:

> I am in one of the most remote, recently opened regions of New Guinea (in the Eastern Highlands), in the center of tribal groups of cannibals only contacted in the last ten years and controlled for five years—still spearing each other as of a few days ago, and only a few weeks ago cooking and feeding the children the body of a kuru case, the disease I am studying. This is a sorcery-induced disease, according to the local people.... It is so astonishing an illness that clinical descriptions can only be read with skepticism; and I was highly skep-

tical until two days ago, when I arrived and began to see the cases on
every side.[14]

Gajdusek and Zigas thus began an intensive investigation under
harsh conditions, documenting kuru's symptoms and trying their best
to make do without proper equipment. By April 3, they already tracked
41 cases.

"The lack of equipment very much restricted our research work,"
Zigas said.[15] Early on, the dining table served as a patient exam table,
lab bench, and autopsy table. The men's food and rum shared space
with enamel plates and wash basins containing brains, organs, and
tissue of kuru victims fixed in alcohol. Soon, a grant from the Australian
public health department enabled buildings to be erected for kuru
research. They were hardly state of the art, with their thatched roofing
and bamboo mat floors, and no electricity or indoor plumbing. Yet the
scientists had a microscope and a host of chemical reagents, enabling
Gajdusek and Zigas to analyze blood, urine, and cerebrospinal fluid.

Part of their task was also epidemiology: assessing the extent of kuru
and the pattern of infection. It might reveal a cluster of cases, suggesting a source of infection or poisoning, or reveal an inherited pattern,
indicating the disease was genetic. One six-day journey to the edge of
South Fore territory "was the most trying experience in my seven years
in the mountainous jungles," Zigas recalled. "Most of the climb of
about 7000 feet was such that we had to ascend hand over hand. Once
attaining the ridge, we then had to descend to 3000 feet, and then
climb another ridge of about 6000 feet; like a yoyo, straight up and
down for long, strenuous hours."[16] Aggressive leeches clasped on; wild
bees disturbed by the movement of their tree-log nests retaliated with
angry stings; swarming mosquitoes encircled their camps; razor-sharp
elephant grass and barbed vines sliced through skin and left gashes ripe
for infection. While such arduous treks left Zigas soaked and breath-
less, forcing him to rest once they reached a village, Gajdusek immediately began interviewing inhabitants and drawing blood.

Tok Pisin is the lingua franca for New Guineans. A pidgin language
drawing from English, German, Spanish, and some native words, it consists of about 2000 terms, making it easy to learn. A natural linguist
who already spoke a half dozen languages, Gajdusek quickly picked up
the tongue, enabling him to establish family histories and chronologies.

Part of the challenge was that the Fore didn't keep track of years—there aren't any seasons on the island, except for dry and rainy. Extensive interviews and connections to known events—such as when the first airplanes were seen—enabled Gajdusek to determine birth orders and estimate ages. From such interviews, the team determined that every village they encountered had had recent kuru cases.

Evenings were a time to review census logs and conduct lab work. It was also a time for Gajdusek to bang furiously on his battered Olympia typewriter, a fascinating draw for the Fore children, who referred to Gajdusek as "Docta America" and "Coutun" (for Carleton). He conveyed his observations regularly to Smadel in the U.S. He compared kuru symptoms to those of Parkinson's disease. Kuru is, he wrote, "a mighty strange syndrome. To see whole groups of well-nourished healthy young adults dancing about, with athetoid tremors which look far more hysterical than organic, is a real sight. But to see them, however, regularly progress to neurological degeneration . . . to death is another matter and cannot be shrugged off."[17]

And so curious to the pioneer bush doctors was the lack of any overt signs of infection—no fever in the patients, no excess white cells in the blood, no inflammation-associated proteins in the cerebrospinal fluid. (In later stages, kuru patients sometimes displayed inflammatory signals, but that came from opportunistic infections that struck as their bodies deteriorated.) Over the next few months, Gajdusek and Zigas sent tissue specimens they collected from around the Fore region to Kainantu for dispatch to Melbourne and to the National Institutes of Health in Bethesda, Maryland, for further analysis.

The key to unlocking the mystery of kuru lay in the brain, and that meant opening up the skull to obtain samples for postmortem analyses. First came the incision of the scalp from ear to ear over the top of the head. Then one flap of skin would be pulled forward over the face, the other to the back of the neck. A single hacksaw blade would take about an hour to cut through the hard skull. Removing the skull cap would reveal the dura mater—the thin, outermost layer of the meninges, the membrane covering the brain. Forceps would be used to raise a bit of the membrane, then a quick slice from a scalpel would part the meninges, exposing the actual brain. Once the frontal lobes and optic nerves were lifted and the remaining nerve connections severed, the jelly-like brain would flop out backward. To solidify the gooey mass so that it could be thinly sliced and viewed under a microscope, the brain would soak for a couple of weeks in formalin, a solution that Gajdusek and Zigas made by mixing 750 milliliters of formaldehyde, 75 grams of salt, and five or six liters of water.

The Fore were cooperative at first and were paid with blankets, axes, and tobacco to bring in their relatives who had died from kuru. "They would barter," Brown recollected. "I asked, for example, if I could take out the liver and the kidney, and they were all, well, you know, how about three trade blankets?" Fore individuals would crowd over Brown as he cut into a body. "There was not this wailing and weeping and grieving and great display of emotion. They were very fatalistic. They knew the disease better than we did, and they knew the lady was going to die and—since they were engaged in cannibalism themselves—they had absolutely nothing against cutting up a body."

Brain Clues

In 1957, Gajdusek managed to send 16 brains to Smadel, who had left Walter Reed and assumed an associate directorship at the National Institutes of Health. There, Smadel assigned the task of examining the specimens to neuropathologist Igor Klatzo, who sliced the brains into thin sections, stained them, and photographed them under the microscope. Not surprisingly, Klatzo reported seeing significant damage to the cerebellum, the part of the brain that controls motor functions. What was surprising was the type of damage. The stained brain tissue revealed vacuoles in the tissue and knots of agglomerated protein— that is, sponge-like holes and amyloid plaques.

To Klatzo, the changes were like nothing he had ever seen before. Certainly no infection, inherited disease, or poison could produce such dramatic pathology. Besides, none of the lab work had isolated anything that could cause such damage. Drawing on a vague memory stamped during his medical training days in Germany, Klatzo thought the pathological features resembled those of Creutzfeldt-Jakob disease. So little was known about CJD at the time that through 1956, only 52 known or suspected cases, under some two dozen synonyms, had ever been reported. Klatzo had to comb through old German journals to find references to it. Yet the differences — the presence of amyloid plaques in about 75 percent of kuru cases, the targeting of mostly women and children, the epidemic spread — clearly indicated that kuru was distinct from CJD. In fact, as it would turn out decades later, it was also distinct from the brain tissue of Stephen Churchill. Related, certainly, but not the same.

Klatzo's photomicrographs were enlarged and became part of a traveling road show about Gajdusek's research on kuru. By the summer of 1959, the images and story found their way to the Wellcome Medical Museum in London. On July 3, a 38-year-old American veterinarian, William J. Hadlow, took the train from Compton, England, to London and sought out the exhibit on the advice of a friend, who had mentioned it over dinner. "It was on the first floor just inside the main door and comprised several panels," Hadlow recalled. "From the start, I was drawn to the neurohistologic changes, especially the vacuolated neurons, unusual in human brains."[18] Hadlow had seen those holes before: not in humans, but in the brains of sheep.

CHAPTER 4

Connecting the Holes

Linking kuru to a disease of sheep enables researchers to experiment with a brain-destroying agent.

The French call it *la tremblante*, or the trembling. In Iceland, it's known as *rida,* meaning ataxia. It's the trotting disease *Traberkrankheit* in Germany; Spaniards refer to it as *prurigo lumbar*, relating it to a skin disorder involving itchy pustules. English speakers call it *scrapie*, because of the tendency of some afflicted sheep to scrape their skin raw.

The fatal neurodegenerative condition comes on quite subtly. It usually takes an experienced shepherd to notice the signs. A sick individual might trail the flock, react strangely to the sheep dog, or become restless. Soon, infected animals may develop intense itchiness and go out of their way to find posts and fences against which to rub. If you scratch a scrapie-infected sheep on the lower back, it may nibble and flick out its tongue, apparently to express satisfaction. The animal may carry its head and ears low, gnash its teeth, nip at its feet and legs, and experience tremors of the head and neck. The gait often becomes wobbly; an affected sheep will sometimes high-step with its forelegs or even "bunny-hop." If you sneak up on an infected sheep and startle it by making a loud noise, it may fall to the ground and convulse. By the time the animal succumbs—usually in one to six months after the onset of symptoms—most of its fleece may have been scratched off, leaving gaping sores.[1]

The place and time of scrapie's origin remain a mystery. It was first described in England in 1732 and in Germany in 1759; by the latter half of the eighteenth century, the disease had begun to spread throughout Europe. Hoping to develop and improve their wool, many European countries had begun to clamor for Spain's famous merino sheep and to breed domestic flocks with them. The imported sheep carried the disease, leading to sudden epidemics within flocks. Inbreeding to improve the wool worsened the outbreaks.

The scourge of scrapie led to eradication programs throughout Europe. Among the earliest directions on what to do with scrapie-infected sheep appeared in Germany in 1759. "A shepherd must isolate such an animal from healthy stock immediately," veterinarian J. G. Leopoldt warned. Moreover, because scrapie is incurable, "the best solution, therefore, is for a shepherd who notices that one of his animals is suffering from scrapie, to dispose of it quickly and slaughter it away from the manorial lands, for consumption by the servants of the nobleman."[2] Evidently, researchers knew that scrapie was contagious among sheep but concluded that they did not pose a health hazard to people—at least to the domestic help.

Still, scrapie persisted despite the culling of sick individuals and, later, the more aggressive measure of total flock depopulation, as veterinarians refer to mass slaughter of potentially exposed animals. New Zealand and Australia are the only two major sheep-raising countries to be free of scrapie, thanks to prompt depopulation of imported sheep that brought the disease down under in 1952. Other countries tried flock depopulation but saw scrapie flare up again once sheep were reintroduced to the area. Iceland, for instance, started its first major push against scrapie and other sheep diseases between 1946 and 1949 by killing all sheep in all areas that ever reported scrapie. Some areas weren't repopulated for three years. Even so, scrapie reemerged within four years of the sheep's return.

An Uncanny Resemblance

William Hadlow had been working for six years as a veterinary pathologist at the National Institute of Health's outpost in Montana—the

Scrapie-infected sheep may itch so much that they scrape off their own fleece. (*Richard Stephenson.*)

Rocky Mountain Laboratories in Hamilton, set up to study the tick-borne Rocky Mountain spotted fever. In the spring of 1958, Hadlow recalled, "I had become restive and readily accepted an offer from the United States Department of Agriculture" to join for three years Britain's Agricultural Research Council Field Station in Compton.[3] (The U.K.'s other center for scrapie research was the Moredun Research Institute, just outside of Edinburgh.) The U.S. had just encountered several outbreaks of scrapie. The first hit in 1947 in Michigan, among Suffolk sheep imported from Canada but originally bred in Britain. Subsequent outbreaks occurred in 1952 and 1954 in California and Ohio. To contain the outbreaks, the Secretary of Agriculture declared a state of emergency. Officials ordered the affected flocks quarantined and depopulated. They also embargoed all British sheep, a move that Hadlow said reinvigorated scrapie research in the U.K.

The U.S. didn't have any domestic centers of scrapie research. So the government had to rely on overseas institutions to teach Americans about the illness. At the Compton lab, scientists were trying to isolate the causative agent and find a line of sheep that resisted scrapie. As a pathologist, "I thought I could best contribute to the British effort by looking at brains, which otherwise were usually discarded," Hadlow said. The main thing researchers knew about the brains of affected sheep were the holes, first reported in 1898 by veterinary researchers Charles Besnoit and C. Morel from Toulouse, France. "Right off, I saw more in them than holes in nerve cells—the well-entrenched diagnostic hallmark of scrapie. Neurons were changed in other ways as well. Many were shrunken and deeply basophilic"—that is, they absorbed stains readily. More impressive, however, was the astrocytosis—the proliferation of the astrocytes, the star-shaped helper cells of neurons.

After a year of watching the disease bring down sheep, examining their brains, and "absorbing the scrapie lore from sundry sources," Hadlow thought he "had a good idea of what scrapie is like: a protracted degenerative disease of the brain, not an inflammatory one, caused by an infectious agent best thought of then as a virus. I did not know of another disease like it in man or animal."[4] That is, until the fateful July day in 1959 when he visited the Wellcome Medical Museum in London to view Gajdusek's chronicle of kuru.

The story was new to Hadlow, who afterward headed to the Royal Society of Medicine Library to track down some of the references cited in the exhibit. Hadlow later wrote:

> I returned to Compton laden with information to mull over in the days ahead. In doing so, I found the overall resemblance of kuru and scrapie to be uncanny. The similarities in epidemiologic features, general clinical pattern, and neurohistologic changes could not be put aside. From these similarities I realized that scrapie might not be unique after all.[5]

And one thing that Hadlow wondered: was kuru transmissible in the same way as scrapie was?

Studying Scrapie

Early on, much debate surrounded the cause of scrapie, which was also found in 1872 to affect domestic goats and in 1992 wild sheep known as mouflons. The disease had characteristics of both an inherited malady and an infectious one. Evidence for an infectious agent came from the fact that scrapie could spread horizontally through a flock—that is, it could move from one individual to another—and that only adult sheep (ranging in age from about two to five years) seemed to be afflicted. Backers of the hereditary view argued that affected sheep showed no fever or other signs of inflammation. There was little pattern to the way the disease spread within flocks—some individual sheep could rub up against sick sheep and remain healthy. And some breeds seem to resist the disease. The early experiments did not resolve matters, either. In 1899, Besnoit reported that he and his colleagues tried to transmit the disease by keeping sick sheep with healthy ones and by transfusing blood and injecting brain matter from affected sheep into normal sheep, but they recorded no successful transmissions.

The thinking about an infectious agent would shift in the early twentieth century to a virus, albeit an unconventional one that took a long time to sicken an animal. In 1934, two other veterinarians from Toulouse, Jean Cuillé and Paul-Louis Chelle, decided to try the transmissibility experiment again. In waiting for his inoculated sheep to show symptoms, Besnoit had watched only for a few months. But Cuillé and Chelle recognized that the incubation period in flocks was more than a year. Suspecting that past experimenters hadn't waited long enough to see the disease, Cuillé and Chelle took a bit of the spinal cord from a scrapie sheep, pureed it to make a homogenous solution, and on July 6, 1934, injected the homogenate into the eye of a healthy ewe. Fifteen months later, the sheep came down with scrapie. This result marked the first experimental transmission of the disease from one animal to another and convincingly demonstrated the presence of an infectious agent. The scrapie agent had made its way along the optic nerve, following it to the back of the brain to the expected location (the right eye leads to the left hemisphere's occipital lobe, the left eye to the right lobe).

The two veterinarians again demonstrated transmissibility when they injected scrapie tissue into the brain and under the skin; they found that the incubation period varied, depending on the technique. It took a year when they inoculated the sheep via the brain and two years when inoculation was done on a peripheral site. In 1939, the two veterinarians reported transmitting scrapie to goats as well.

Poking a syringe into an animal's brain to give it a lethal neurological disease may seem cruel, but such work was critical for understanding the illness. It meant that scientists could study the incubation period, the symptoms, and the effect of the individual's genetics on the course of the disease. Goats proved to be better lab animals than sheep because they were genetically more susceptible, almost uniformly coming down with the disease once inoculated. In sheep, you were lucky if a third developed it—the Cheviot sheep breed at Compton came down with it 25 percent of the time after inoculation. It was not until 1961, when Compton's Richard L. Chandler discovered that it was possible to transmit the scrapie agent to a much cheaper and more prolific animal: mice. Scrapie researchers could trade the barn and grass for cages and shredded newspaper.

Trying Transmissions

Hadlow's recognition of the neuropathological similarity between kuru and scrapie led him to propose a transmissibility experiment. He wrote to the British journal *The Lancet,* but fearing delayed publication because of a printer's strike, he also sent a letter to Carleton Gajdusek. "I've been concerned primarily with the syndrome induced experimentally in the goat by intracerebral or subcutaneous inoculation of brain tissue from scrapie-affected sheep," he wrote on July 21, 1959. "The lesions in the goat seem to be remarkably like those described for Kuru. . . . All this has suggested to me that an experimental approach similar to that adopted for scrapie might prove to be extremely fruitful in the case of Kuru."[6] Hadlow was proposing that Gajdusek try inoculating bits of kuru brain into healthy brains—not those of humans, of course, but those of other primates.

Gajdusek hadn't heard of scrapie, but not wanting to let on, he wrote back to Hadlow on August 6, 1959, and gave Hadlow the impression that such transmissibility experiments were actually underway. The two met face-to-face for the first time later that year, on the first stop of Hadlow's scrapie-information tour of the U.S. The tour, which began on November 23, 1959, in Washington, DC, was designed to ease the minds of American sheep men. "That was when I first met Carleton Gajdusek, the young man with a crew cut who stood silently in the back of the room while I gave my talk. After the meeting, he came up to me and introduced himself. I am sure we talked about my letter, but I have no recollections of what was said," Hadlow stated.[7]

Gajdusek returned to New Guinea in early 1960 to conduct more bush patrols to find kuru victims. Later that year, he visited scrapie research labs in the U.K. and Iceland, and by winter he was convinced of the importance of transmissibility experiments on animals, especially chimpanzees, the closest relatives of humans. In 1961, Gajdusek and his colleague J. Anthony Morris obtained permission to use the secluded Patuxent Wildlife Research Center, spread over 5000 acres in Laurel, Maryland, for the inoculation studies of kuru. A small, uninsulated cinder-block building would be constructed to house the chimps and monkeys, a task that would not be completed for two years.

The notion of a slow-going lab experiment—it might take years for the animals to come down with the disease—didn't suit the peripatetic

Gajdusek, who liked to be away from his National Institutes of Health office in Bethesda, Maryland, for months at a time. So in the summer of 1961, he and Joseph Smadel, who a year earlier had become chief of the NIH Laboratory of Virology and Rickettsiology, tried to recruit Hadlow to monitor the inoculation of the chimps. "I declined their offer, concluding, unfairly as it turned out, that anyone who took the job would become little more than an exalted handler of apes," Hadlow recalled.[8] Besides, he wanted to continue his scrapie research back at the NIH's Rocky Mountain Laboratories.

Smadel turned to Clarence J. "Joe" Gibbs, a former protégé of his at Walter Reed. At the time, Gibbs was thinking of taking a Rockefeller Foundation fellowship to study mosquito-borne viruses in South America and had sought Smadel's advice. "Smadel's reaction was immediate and violent and in his inimitable fashion he pointed his finger in my face and said, 'Goddam it Gibbs you're not going to Brazil!'" Smadel then told Gibbs that the Rockefeller Foundation was about to pull the plug on overseas research. Gibbs asked Smadel where he should go. "His reply was that I was going to the Patuxent Wildlife Research Center, Laurel, Maryland, where I would study scrapie disease of sheep and attempt to transmit to chimpanzees" and other animals.[9] The center would eventually house dozens of chimpanzees and many more monkeys, thousands of mice, and other animals for the transmission experiments.

With Gibbs readying the laboratory, Gajdusek found some help in dealing with the human face of kuru in New Guinea. Australian officials assigned a young Adelaide physician from the public health department, Michael P. Alpers, to the area in late 1961. With his wife and baby daughter in tow, Alpers joined Gajdusek. With his gracious, thoughtful, and taciturn ways, Alpers proved to be an ideal complement to the irrepressible and voluble Gajdusek. The two continued with the patrols, enduring long hikes through rugged terrain—at one point, they unwittingly walked into the crater of an active volcano burbling with water infused with hydrogen sulfide gas. They mapped out the extent of kuru and retrieved brain, blood, and other parts of kuru victims to send to the Hall Institute in Melbourne, where the samples were inoculated into lab animals and chicken embryos, and to Joe Gibbs at the NIH.

Because none of the early transmission experiments with kuru were successful, "one possibility we thought for the lack of transmission was that [the kuru agent] was dying before we could get it back here from

New Guinea," said Paul Brown, who followed Alpers as kuru bush pathologist after Alpers joined Gibbs at the NIH in 1964. "So I camped out in the huts, in a little village where there would be people dying. I went with all this elaborate network of liquid nitrogen."[10] Brown would draw from these great vats of the supercold liquid (–196° C) so he could immediately freeze the brains and organs he removed. At the time, none of them knew that such measures were quite unnecessary.

Georgette's Sacrifice

The real trick for successful transmission was to hold onto the animals for many months or years, as Hadlow had suggested in his *Lancet* letter, and not the typical two to three months, as other infectious disease specialists did in their experiments. With a dedicated facility at Patuxent, holding onto a variety of animals for lengthy periods became possible. The mice that Gajdusek and Morris had previously inoculated with scrapie at their NIH Bethesda lab in August 1961 were transferred there, and these rodents were soon joined by a small zoo.

The inoculation of primates began in August 1963. The chimps received the usual dispassionate designations—A1, A2, and so forth—but the researchers also gave them names like Daisy and George (who later upon maturation was discovered to be Georgette). Daisy was the first to be inoculated. After knocking the chimp out with a whiff of ether, the scientists drilled a little hole in her forehead. They injected into the left frontal cortex 0.2 milliliter of a solution consisting of 10 percent brain matter from a kuru victim named Kigea. Georgette got her dose from kuru patient Enage. "Within minutes following the inoculation, the chimpanzees were once again roaming the laboratory and sitting on the secretaries' desks," Gibbs recalled.[11] "All in all, by the end of 1963 I had inoculated about 10,000 mice, 7 chimpanzees, and 75 smaller non-human primates." Along with a technician, Gibbs cared for the animals and waited for signs of kuru-like symptoms. He relied on letters and cablegrams to keep in touch with Gajdusek, by now a part-time New Guinean.

"I had never seen a patient with kuru nor had I reviewed the many hundreds of feet of cinema film . . . of kuru patients and thus I was not at all sure that both chimpanzees were responding to their inoculations or whether they had merely developed intercurrent infections," Gibbs wrote.[12] It was his lab assistant who actually first noted signs in June 1965. Georgette had the "shakes," her lower lip drooped, she fell off the top of the cage. Soon, she would no longer reach for the food that was right in front of her. She would stoop down, her arms to her sides, and reel in the food with her mouth—a peculiarity Gajdusek later referred to as a "vacuum cleaner"–style of feeding. Daisy, too, was beginning to exhibit similar signs. "It soon became evident that the symptoms were slowly progressive" and not associated with any known acute infections, Gibbs concluded.[13] To Michael Alpers, who had seen plenty of kuru victims in New Guinea, the signs were unmistakable.

In a few months, Gibbs had to establish a daytime nursing schedule to care for the deteriorating apes; they had become so ill they had to be hand-fed. Eventually, the chimps required round-the-clock attention. Gibbs even brought in "a plethora of nationally and internationally recognized neurologists" to look at Daisy and Georgette. "No human patient would have received the medical attention these animals received. . . . The superb nursing care . . . allowed us to study the progressive course of the disease over several months with no apparent discomfort to the animals," said Gibbs, who with his staff had grown quite enchanted with the remarkable animals. By the end of October 1965, the team decided it was time: they anesthetized Georgette and drained her blood. A few months later, in February, Daisy would also be sacrificed. "Their loss, even in the establishment of a remarkable scientific event, was felt by the staff," Gibbs wrote.[14]

For Georgette's postmortem, Gajdusek recruited Elizabeth Beck, a neuropathologist from London who had worked extensively with scrapie brains. Gibbs, Alpers, and Beck took everything they could out of the body, freezing a bit of brain for viral analyses and fixing the rest in formaldehyde. Organs including the liver, spleen, and kidney were saved in jars—even the arms and legs were twisted out of their sockets, skinned, and preserved. After taking Georgette's brain back with her to London, Beck waited three weeks for the fixation to be completed. Then she sectioned the brain into thin slices suitable for viewing under the microscope. In December 1965, she reported to Gibbs and Alpers: Georgette's brain looked just like those of human kuru victims.

The Kuru–CJD Link

Over the next few years, Gajdusek, Gibbs, and Alpers continued with their inoculations, injecting brain homogenates from different kuru victims into chimps and showing experimental transmissibility. To prove that chimpanzee brains incubated the kuru agent, the team had to demonstrate serial passage—healthy chimps inoculated with homogenates from the brains of sick chimps should come down with the disease. Sure enough, Daisy's brain proved to do to chimps what the brain of kuru victim Kigea did to Daisy. Later, the scientists showed that the disease could be transmitted to monkeys from the New World (capuchin, marmoset, spider, squirrel, and woolly) and Old World (green, bonnet, cynomolgus macaque, mangabey, pig-tailed macaque, and rhesus). "Thus, we had established 'slow virus infections' as a cause of subacute progressive degenerative disease of the central nervous system of man," Gibbs concluded.[15] The kuru agent proved quite infectious—one gram of infected brain had the potential to kill 100 million lab animals.

Kuru wasn't the only disease Gajdusek wanted to try transmitting to lab animals. He was curious to see if amyotrophic lateral sclerosis (ALS), multiple sclerosis, Alzheimer's, Parkinson's, Pick's, and other neurological diseases were similarly transmissible. His team prepared homogenates from the brains of people who had died from these diseases and injected them into test animals. Only one disorder proved to be transmissible, and it came as no surprise: Creutzfeldt-Jakob disease. Igor Klatzo had already noted the similarity in damage to brains from kuru and CJD patients. Over the next several years, Gajdusek and Gibbs would show that brain tissue from more than a dozen CJD patients could transmit the disease to chimps, monkeys, and several non-primate species such as cats, ferrets, guinea pigs, mice, and hamsters. The incubation periods for chimps inoculated with CJD was around 10 to 14 months, compared with the 14 to 39 months for kuru. (On second passage, from chimp to chimp, the incubation period dropped to 10 to 12 months.) "The basic cellular lesion, best appreciated by electron microscopic examination," Gadjusek explained, "is the same in both kuru and Creutzfeldt-Jakob disease: a progressive vacuolation"—a formation of holes within the dendrites and axons of the neurons, leading to the eventual destruction of the cell.[16] The symptoms and other clinical features of neurodegenerative diseases typically overlap. But kuru and CJD stand out with the spongy holes they leave in victims' brains and their transmissibility in the lab. Here, then, was a new category of disease: the transmissible spongiform encephalopathies, or TSEs.

An End to an Epidemic

Here, too, was evidence for what "everybody" already knew about kuru: The rituals of cannibalism spread the disease. In the late 1960s, anthropologists Shirley Lindenbaum and Robert Glasse presented persuasive evidence of the cannibalism–kuru connection—showing, for instance, that kuru arose around the same time as cannibalism began.

Based on the lab experiments, Gajdusek concluded that it probably wasn't the actual consumption of contaminated human flesh that brought on kuru—chimps didn't contract the disease when they were fed the infected tissue. More likely, infection occurred as a result of handling the diseased matter. The Fore, Gajdusek described, "did the autopsies bare-handed and did not wash thereafter; they wiped their hands on their bodies and in their hair, picked sores, scratched insect bites, wiped their infants' eyes, and cleaned their noses, and they ate with their hands."[17] (In 1980, however, Gajdusek's team did find that squirrel monkeys could get kuru via oral consumption, but less efficiently than by the intracerebral route.) The epidemic probably began when a Fore individual came down with the sporadic form of Creutzfeldt-Jakob disease, which occurs spontaneously in one of every million people.

The Fore's story ends well. Although the tribespeople worried throughout the 1960s that kuru—and the revenge killings of supposed sorcerers who had allegedly created it—might mark their doom, the seeds for the solution to the kuru problem had been planted. Embarrassed by what they perceived as savagery, Western missionaries and the Australian administration managed in the late 1950s to persuade the Fore—sometimes through means of police arrest—to abandon cannibalism. Soon, a curious trend began to shape up in the epidemiology of the disease: As the 1960s wore on, the age at which the youngest victims came down with kuru was increasing. In fact, no one born after 1959 contracted kuru. The cessation of cannibalism meant the end of new kuru infections.

Michael Alpers, who in 1977 began his service as the director of the Papua New Guinea Institute of Medical Research, thought he would be around to see the last case of kuru. Surprisingly, however, kuru has yet to become an extinct disease. By the time Alpers retired in 2000, one or two cases were still cropping up on occasion; the latest one occurred in early 2003. Victims fall in the middle-aged and elderly categories, born before the end of cannibalism. Evidently, the incubation period of kuru can exceed 40 years.

Nobel Worthy

Joe Gibbs recalled that as Elizabeth Beck boarded the plane to London with Georgette's brain, she predicted that a Nobel Prize would come from the studies.[18] She was right. In 1976, the Royal Swedish Society in Stockholm awarded Carleton Gajdusek the Nobel in Physiology or Medicine, for "discoveries concerning new mechanisms for the origin and dissemination of infectious diseases." (He shared the prize with American biochemist Baruch S. Blumberg, who found the antigen for the hepatitis B virus.)

Gajdusek fretted that Vincent Zigas and Joe Gibbs were not included (the prize can be shared by no more than three people), but he believed there was another Nobel in this field. After all, the *thing* that actually caused kuru and CJD still eluded researchers' grasp. Gajdusek and his colleagues had managed to prove that some sort of infectious agent was present in bits of brain, but they did not isolate it. Discovering the agent—still presumed to be a slow-acting virus—and characterizing its structure, Gajdusek felt, would surely warrant one of those famous early-morning calls from Stockholm. Like Beck, Gajdusek was prescient—another prize would be awarded in this field, but not for a virus.

CHAPTER 5

The Birth
of the Prion

*The unusual mode of attack and biochemical durability of
the TSE agent leads to an heretical idea.*

That the transmissible spongiform encephalopathies such as Creutzfeldt-Jakob disease, kuru, and scrapie provoked no overt immune
response in their victims — no fever or swelling or chills — was the first
clue that the causative agent was like nothing ever encountered before.
All disease-causing entities — parasites, fungi, bacteria, and viruses —
indicate their presence in some way, however subtly.

As generally defined today, parasites are tiny animals that rely on our
bodies for food, shelter, and transportation. They could be one-cell pro-
tozoa such as the plasmodium parasite, which causes malaria, or the
trypanosome, which causes the deadly sleeping sickness in Africa.
Others are bigger, multicellular creatures such as helminths — worms
and flukes, some of which can grow to several feet in length. The
immune system typically mounts an initial attack on parasites, but
many have evolved strategies to survive the onslaught. Consider the 20-
micrometer-long trypanosome, which reproduces by splitting in two
about once every eight hours. The trypanosome progeny adopts a
slightly different surface chemistry to confuse our bodies. Because subsequent generations of trypanosomes don slightly different coats, the
immune system has to re-recognize the parasite again and again.
Staying a step ahead, the trypanosomes can persist for years in the

bloodstream, causing cycles of fever and headaches as the body tries to fend them off. Eventually, the trypanosomes make their way to the nervous system, where they ultimately induce coma and death.

Infections from fungi (molds and yeast) are familiar to many people. In athlete's foot, the fungus infects the skin cells and produces itching and scaling. More seriously, fungi can invade the internal organs and form colonies sufficiently large to block off the bronchial tubes, choke off arterial valves, or plug up the ureters leading from the kidneys. Blastomycosis, cryptococcosis, and candidiasis are some of the more serious fungal diseases.

Bacteria—single-celled creatures—are the most abundant life-form on earth. The large intestine alone harbors 10 trillion to 100 trillion of them, exceeding the number of cells in the human body. They average about 1 micrometer wide, and 100 million can fit on the pinky nail. Most bacteria are harmless, and many are, in fact, crucial for health, as they break down food in the gut that would otherwise be indigestible. There are plenty, of course, that cause disease; even strains of otherwise beneficial bacteria, differing only slightly in genetic makeup, can be deadly. Some bacteria, such as *Bacillus anthracis* (anthrax) and *Clostridium botulinum* (botulism), produce powerful toxins that kill. Other bacteria kill indirectly by stimulating an overreacting immune response that ultimately damages vital organs with, for example, blood clots in the case of sepsis, or fibrous tissue deposits in arteries leading to multiple strokes in the case of tuberculosis.

Viruses are unique—their reason for invasion is not quite the same as those for parasites, fungi, and bacteria, all of which look to our bodies as a source of nourishment and a safe place to reproduce. Viruses are really no more than strands of genetic material—either DNA or RNA, never both—surrounded by a protective coat made of protein. So meager are the instructions encoded by viral nucleic acids that viruses cannot reproduce on their own. Because a virus can't replicate by splitting in half as other microorganisms do, it needs to take over a cell. The virus forces it to follow directions as written in the viral genetic code. The code tells the enslaved cell to create components, such as nucleic acids or proteins, which the virus needs to make copies of itself. The virus progeny eventually leaves the host cell, either by budding off the cell or simply working the cell to death so that the cell breaks apart and spills its contents. The freed viruses can go on to infect other cells or leave the body to infect the cells of other organisms.

DNA MAKES RNA MAKES PROTEIN MAKES LIFE

Contained in virtually every cell of your body is your genome—the complete set of instructions that tells cells what to do, how much of it to do, and when to do it. It also tells the cell when it's time for mitosis—the process of dividing into two daughter cells, each of which takes a copy of the genome, ensuring that subsequent generations of cells will have the same instructions. Egg and sperm cells are different: they have half the amount of genetic information present in other cells. During fertilization, the sperm and egg fuse, giving the resulting embryo the full complement of genes—and allowing you to have your mother's nose and your father's eyes.

The Genetic Code

The genetic instructions are encoded by a long molecule resembling a twisted ladder called deoxyribonucleic acid—DNA. Remarkably, all of DNA's information is encoded by nucleotides, its basic constituents. Nucleotides consist of sugar (called deoxyribose), phosphate, and one of four bases: adenine, thymine, cytosine, and guanine— better known by their initials A, T, C, G. The sugar and phosphates join together in a line to form a backbone that holds the bases. Pairs of strands are joined together by pairs of bases that match up in specific ways. Adenine in one strand always pairs with thymine in the other strand, and cytosine pairs with guanine. Latched together, the base pairs form the "rungs" of the DNA ladder. The twisting of the ladder yields the famous double-helix shape of DNA, which James Watson and Francis Crick identified in 1953. Inside the cell nucleus, the double helix stays tightly coiled up—there's about 6 feet worth of DNA in each cell. The DNA bundles up into structures called chromosomes, which are visible under the microscope when a cell is about to divide. Humans have 23 pairs of chromosomes; 22 pairs are the same in males and females and are called autosomal chromosomes. The remaining pair are the sex chromosomes, designated X and Y. Females have two X chromosomes, and males have one X and one Y.

The genetic code is made up of the four-letter alphabet A, T, C, and G. Combined in certain sequences, the bases form genes just as letters form words. For example, the genetic sequence of bases on one strand might look something like this: ACC-CCAGCTGTTGGGGCCAG. The example shows the first 20 bases of the code that explains how to make myoglobin, a substance that delivers oxygen to muscles. The rest of the code for myoglobin has another 1000 bases. Human chromosomes have about 3 billion base pairs that spell out an estimated 30,000 to 40,000 genes.

Making Proteins

All these genes are instructions for making proteins. To many people, proteins are most familiar as an essential part of a balanced diet, but proteins are also the materials that carry out the activities of life. They make up the structures of cells and the compounds needed for cells to do their jobs. They are the enzymes that digest food, the substances that relay signals between nerve cells, the hormones that regulate metabolism, and the clotting factors that seal ruptured blood vessels.

The cells of your body are kitchens designed to cook up various kinds of proteins. And just as you need to go to the supermarket to stock the kitchen, so, too, do your cells need the raw material from food to make proteins. The carbohydrates, fats, and sugars you consume provide the energy to make those proteins.

When it's time for cells to make proteins, the DNA strands unzip to expose the A, T, C, and G bases. Nucleotides floating freely in the cell nucleus then bind to them to form messenger RNA, or ribonucleic acid (the process of transferring the code from DNA to RNA is called transcription). Since both are nucleic acids, RNA and DNA are very similar; however, RNA has a slightly different chemical structure and relies on the base uracil rather than thymine to bind to adenine. These chemical differences make RNA single-stranded and enable it to carry out its functions.

As its name suggests, messenger RNA does the work of making deliveries, taking the DNA instructions out of the nucleus and into the cytoplasm. Why doesn't the DNA just carry the instructions out itself? Just as a valuable recipe might be kept in a drawer to protect it from spills and splatters, the DNA stays safely tucked in the nucleus of the cell to protect itself from the chemically active environment of the cytoplasm—it's the master copy, after all. Other reasons for relying on messenger RNA have to do with efficiency and flexibility in protein manufacture under changing conditions: For example, if a cell is cold (or hot or starved) and needs a surge of a particular stress protein for protection, then having multiple copies of RNA is a quicker way of making a lot of the protein than it would be by having a single copy of DNA do it.

Once outside the nucleus, the messenger RNA takes the instructions to a ribosome, where the proteins are made. The messenger RNA is read, and protein-manufacturing begins (the process is called translation). Proteins are actually blocks of amino acids, which are derived from food and float around inside cells. Each amino acid consists of ten to about two dozen atoms—mostly carbon, hydrogen, oxygen, and nitrogen.

In the ribosome, the genetic code on the messenger RNA is read three letters at a time. These three letters are called a triplet or a codon. The four letters A, T, C, G can form sixty-four possible three-letter combinations. Sixty-one of these combinations code for amino acids. For example, the codon adenine-uracil-guanine in RNA (adenine-thymine-guanine in DNA) specifies the amino acid methionine. The other three are called stop codons and tell the ribosome to put the brakes on protein manufacture. The human body uses just twenty different amino acids; most amino acids have more than one codon. Amino acids will keep linking together in the ribosome until a stop codon is encountered. It then releases the completed chain—a protein.

The protein is not ready to go to work just yet. In a manner that is still not completely understood, slight intermolecular forces—one amino acid pulling away or pushing toward another amino acid—cause the protein to fold up into a tight bundle. Only proteins that have been properly folded can function in the body. The cell generally breaks apart any misfolded proteins and recycles their amino acids.

The whole protein-making process occurs in all living cells. It can be remembered rather simply: DNA makes RNA makes protein.

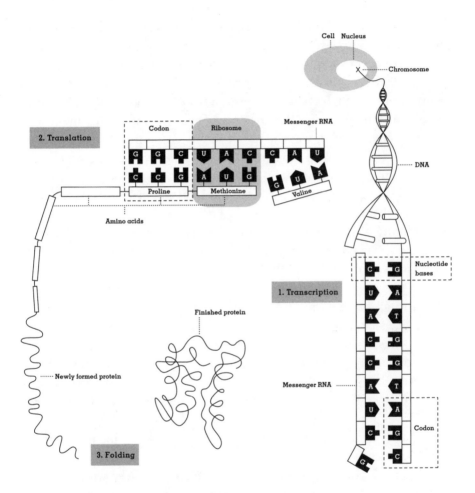

Protein-making in cells begins when DNA, coiled up in the chromosome, unwinds to reveal its bases: adenine (A), cytosine (C), thymine (T), and guanine (G). During *transcription* (1), the DNA code is transferred to messenger RNA (instead of T, RNA relies on uracil, or U). The messenger RNA is then held at the ribosome while it dictates the order in which amino acids are linked, a process called *translation* (2). It takes three bases (called a codon) to specify an amino acid. The linked amino acids form a stringy protein, which has to fold up into a complex shape before it can function (3).

A Tough Invader

By the mid-1960s, most scientists studying scrapie, kuru, and other transmissible spongiform encephalopathies (TSEs) believed that viruses were behind the afflictions. One clue was the simply size—most bacteria and parasites are big enough to be seen through a standard microscope, yet samples from TSE-infected tissue yielded no relevant microorganism. Viruses are far tinier than bacteria and range in size from about 0.02 to 0.25 micrometer; most can only be visualized with an electron microscope. The real proof that the infectious agent was virus-sized or smaller was the use of filters with pores sufficiently fine to block bacteria and anything larger. Carleton Gajdusek and Joe Gibbs, for example, strained CJD-infected tissue samples through pores 0.22 micrometer wide; the strained material still proved infective, showing that the agent passed through the filter.

The tiny TSE pathogen proved to be exceedingly durable. The first indication of its toughness was revealed by accident in the 1930s. William S. Gordon, the then director of the Compton lab in England, developed a vaccine to treat a sheep disease called louping ("looping") ill, an affliction that was first described in the nineteenth century in the U.K. The sickness, caused by a tick-borne virus, ravages the brains of sheep, causing them to jump as they walk, a gait the British call louping. To produce the vaccine, Gordon took bits of brain and spleen from infected sheep, homogenized them, and poured in a bit of formaldehyde to inactivate the virus. When the vaccine is injected, the inactive virus stimulates a response from the immune system. So trained, the immune system would in the future quickly attack any incursions from the live virus.

In 1935, Gordon began inoculating some 40,000 sheep with the louping-ill vaccine, successfully protecting the animals from the viral disease. Unfortunately, in 1937, some of the sheep began coming down with scrapie. To his horror, Gordon realized that one of his vaccine batches was contaminated with scrapie, and he had given it to some 18,000 sheep. About 1500 sheep ultimately got scrapie from the vaccine; the figure probably wasn't higher because many of the sheep had been adults at the time of inoculation and were slaughtered before symptoms appeared. Besides being an unintended transmission experiment, Gordon's vaccination program showed how tough the scrapie agent is.

The sheep tissue used in the preparation had soaked in formaldehyde, a toxic preservative that readily destroys viruses and other microorganisms. Yet the scrapie agent survived this potent chemical.

Soon, other unusual properties of the agent became apparent. Freeze the agent, thaw it out, and it will be none the worse for wear. For a bigger surprise, put a bit of infected tissue into water and bring it to a rolling boil. After a few minutes, all bacteria and parasites will be killed. But the scrapie agent will retain its deadliness. Its resistance to dry heat is even more impressive: The agent remains infective even after cooking at 600°C. That would surely make for one unconventional virus.

The agent's ability to survive extreme temperatures, however, pales in comparison to its capacity to withstand the assault leveled against it by Tikvah Alper of the Hammersmith Hospital in London. A native of Cape Town, South Africa, she studied the effects of radiation on cells and various organisms, including bacteriophages (viruses that infect bacteria). In 1966, Alper, working with David A. Haig and Michael C. Clark from the Compton laboratory, took some dried scrapie-infected brain tissue and bombarded it with high-energy beams of electrons. Such electrons have enough energy to knock other electrons from atoms in the sample, thereby creating positive and negative ions — hence, the beams are referred to as ionizing radiation. Sufficiently energetic beams can destroy cells. By blasting the scrapie-infected tissue, Alper and her colleagues wanted to determine the dose needed to inactivate the agent. From there, they could calculate its size: the more intense the beam, the smaller the target molecules that can be ionized. "We concluded that the agent was smaller, perhaps by a factor of 10, than any known virus," Alper recalled.[1]

More curious results would appear a year later, in 1967, when the Alper team conducted a follow-up experiment. They mixed scrapie-infected brain tissue in water and exposed it to the ultraviolet light of a low-pressure mercury lamp. The wavelength of such light, 254 nanometers, causes nucleic acids to break apart. Fracturing DNA is an efficient way to sterilize air, which is why hospitals and high-tech air filters use the lamps. Exposing the scrapie agent to this germicidal light, however, did nothing. "We could discern no inactivating effect of doses that were enormous compared with those required to destroy the function of any known nucleic acid entity that had been tested up to that time," Alper explained. The ionizing radiation and ultraviolet light experiments, taken together, "led us to moot the possibility that the agent has a mode of replication independent of the integrity of a nucleic acid moiety."[2] In other words, the scrapie agent didn't need genes to function—which is like saying a skyscraper can go up without any blueprints.

The Elusive Agent

In the wake of the irradiation reports, scientists began speculating furiously as to the nature of the scrapie agent. Biochemical attempts to isolate the agent had failed. "A prolonged period of intense frustration ensued," recalled biochemist and noted scrapie researcher Gordon Hunter of the Compton lab. "Try how we may, using solvents, enzymes, detergents and other chaotropic agents, it proved impossible to separate scrapie activity from membranous components"[3]—the proteins, lipids, and other cellular materials normally present in the brain. Examining scrapie brain slices under the electron microscope, the most powerful imaging tool at the time, revealed nothing. So investigators began tossing out ideas—perhaps it was a bit of DNA with a polysaccharide (carbohydrate) coat, or a parasite resembling the tiny, muscle-infesting sarcosporida, or a small DNA virus, or a small virus that could generate RNA particles. Maybe it could survive ultraviolet blasts because, like a mouse pathogen called the polyoma virus, it can repair its DNA. Maybe the scrapie agent is phlogiston, linoleum, or kryptonite, some jested. By 1975, the hypotheses would outnumber the experimental groups studying TSEs.

Kicking around in the background was the sacrilegious idea that the agent was just a protein. Hunter explained that the first mention along those lines occurred in 1959, by John Stamp of the Moredun Research Institute near Edinburgh, Scotland. Later, Iain Pattison, another Compton scrapie scientist, "was particularly impressed with the resistance of the scrapie agents to reagents such as formalin," Hunter said, "and in a heretical paper he emphasized his view that scrapie could not be classed with conventional viruses. Like so many others, however, he overstated his case."[4] In a 1967 paper, Pattison had hypothesized that the scrapie agent was a basic protein.[5] But in forming his conclusion, Hunter explained, Pattison had relied on experiments where cross-contamination was not controlled; moreover, the basic proteins isolated from scrapie brain were not biologically active whatsoever.

A conceptual breakthrough came from mathematician J. S. Griffith from Bedford College in London. In his 1967 *Nature* paper, he proposed, on theoretical grounds, how proteins could replicate without nucleic acid (DNA or RNA).[6] He pointed out that, at least in terms of the laws of physics, it was possible for the scrapie agent to be a malformed version of a normal protein existing in healthy hosts. The agent could in principle serve as a template that created equally malformed versions of the host's protein and hence lead to disease. But "there were at the time no ways of testing for the transmission of information back from protein into nucleic acid to reverse the functional direction of the genetic code," Hunter noted. "It was really pure speculation rather than a hypothesis."[7]

In a time that Hunter refers to as "the period of false trails," the 1970s witnessed several claims of important finds, but none panned out. Scientists in the U.S. and the U.K. said they detected blood factors essential to the development of scrapie. Several researchers claimed to have found the agent and named it after themselves, such as Cho particles or Narang particles. Other reports proclaimed to have spied nucleic acids unique to scrapie brains. All these leads proved false — they either resulted from contamination, or were not specific to scrapie-affected brains, or simply proved to be irreproducible by other experimenters.

TSEs' New Player

The period of false trails may have been frustrating to TSE investigators, but it was an ideal time for new blood to enter the field. Knowledge was so limited that there were plenty of territories that an ambitious, talented researcher could stake out in the hopes of a major contribution.

While Carleton Gajdusek was off collecting kuru brain samples and shipping them to the U.S., Stanley B. Prusiner was busy with his algebra, Latin, science, and other homework assignments from Walnut Hills High School in Cincinnati. Prusiner was born in Des Moines in 1942, but in 1952, his father, an architect, relocated the family to Cincinnati for good. Prusiner decided to attend college away from home, heading east to the University of Pennsylvania, where he majored in chemistry and later finished medical school. He spent three years at the National Institutes of Health, where he honed his scientific research skills by studying the enzymes used by the bacterium *E. coli*. "As the end of my time at the NIH began to near," Prusiner recounted, "I examined post-doctoral fellowships in neurobiology but decided a residency in Neurology was a better route to developing a rewarding career in research. The residency offered me an opportunity to learn about both the normal and abnormal nervous system."[8] He began his residency in neurology at the University of California, San Francisco, School of Medicine in July 1972.

His life changed completely just two months later, after he admitted an elderly woman suffering from progressive memory loss and motor coordination problems. The 30-year-old Prusiner was surprised to discover she had Creutzfeldt-Jakob disease. He quickly began to learn as much as he could about TSEs, which wasn't much at the time. "The amazing properties of the presumed causative 'slow virus' captivated my imagination and I began to think that defining the molecular structure of this elusive agent might be a wonderful research project. The more that I read about CJD and the seemingly related diseases—kuru of the Fore people of New Guinea and scrapie of sheep—the more captivated I became," Prusiner recalled. "Over the next two years I completed an abbreviated residency while reading every paper that I could find about slow virus diseases."[9]

Because he lacked training as a virologist and had not worked with anyone involved in the field, the National Institutes of Health rejected Prusiner's first grant proposal for a scrapie study. So Prusiner set up a collaboration with William Hadlow and Carl M. Eklund from the NIH's Rocky Mountain Laboratories. Prusiner was dead-set on isolating the scrapie agent, even though colleagues warned him about the high-risk nature of the work—specifically, the tedious, laborious, and expensive assay methods.

To isolate a causative agent, virus hunters rely on a process called end-point titration. Take the sample, dilute it by half with a buffering agent to keep the pH stable, and spin it in a centrifuge to separate the heavier components from the lighter ones. Then pipette out a bit of the sediment and a bit of the supernatant (the usually clear liquid above the sediment) and test each sample by injecting it into healthy hosts. Wait for results. The fraction that kills the host the fastest is the purest fraction—that is, the portion with the most infectious agent and the least amount of extraneous material. Repeat with progressively more watered-down solutions until you find the most dilute fraction that can still cause disease. That gives you a measure of the viral concentration in the original sample.

The problem for scrapie workers was the long incubation periods—sheep took years to come down with symptoms. Mice were a vast improvement, but the work still proceeded slowly. The most concentrated samples produced symptoms in four to five months; the weakest samples might take more than a year. Dilute, spin, inject, wait, repeat. Dilute, spin, inject, wait, repeat. About ten dilutions were necessary. Prusiner later estimated that it might take hundreds of thousands of mice and a few lifetimes. "We rapidly went through our ten thousand mice, and even if we were handed money on a silver platter, we couldn't go on like that," Prusiner told science journalist Gary Taubes in a 1986 *Discover* magazine article.[10]

In 1978, Prusiner and his colleagues found an alternative to endpoint titration. Rather than using mice, they decided to go with hamsters. In 1975, Richard Marsh of the University of Wisconsin at Madison and Richard Kimberlin, then of the Moredun Research Institute near Edinburgh, Scotland, discovered that hamsters came down with a form of scrapie twice as quickly as mice. Moreover, Prusiner found strong correlations between the concentration of scrapie agent and the rapidity of both disease onset and death. "Thus instead of determining how much a sample could be diluted and still cause disease, we measured how fast a sample with a known dilution brought on disease symptoms and caused death," he wrote in his October 1984 *Scientific American* article.[11] The switch to hamsters and the incubation-time assay accelerated research 100-fold. "Instead of observing 60 animals for a year, we can assay a sample with just four animals in 60 days," stated Prusiner, who concluded that his team conducted more experiments on the biochemistry of scrapie in roughly two or three years than anyone had ever done in the entire history of scrapie research.[12]

By 1981, Prusiner had set up his own lab at UCSF and, with his incubation-time assay, managed to achieve an overall purification factor of

100—his preparations were as infectious as brain samples but were made of 99 percent pure scrapie. Along the way, his team discovered that the scrapie agent varied quite a bit in size and density. Based on how fast the agent settled during centrifugation, the infectious particles could be smaller than the smallest known viruses, or as big as mitochondria or bacteria. Evidently, the agent could clump into differently sized clusters.

Biochemically, the agent remained consistent with Tikvah Alper's irradiation work. When Prusiner added compounds that destroy or modify nucleic acids—such as nucleases, zinc ions, or hydroxylamine—the samples remained infective. But when he treated the agent with substances that denatured or digested proteins—unraveling their folded structures or cutting their amino acids apart—the samples lost their ability to induce scrapie.

Prusiner also developed another line of argument to support the notion that the agent was a protein. Alper's work already put size limits on the agent, indicating that the molecular weight was between 60,000 and 150,000 daltons. (One dalton, or atomic mass unit, weighs about as much as a hydrogen atom, or about 1.66×10^{-24} gram.) Prusiner's lab put the agent through several other kinds of tests to determine its size, seeing if it fell through membrane filters of known dimensions and racing it through variably dense substances. He lowered the estimated molecular weight to 50,000 to 100,000 daltons—the agent could be only 5 nanometers wide, 1/100 the size of the smallest known viruses.

Of course, some nucleic acid could still lurk within the protein shell, but it would have to be a snippet—maybe a dozen to fifty nucleotides long. The standard genetic code requires three nucleotides to specify an amino acid, so the protein could only be composed of a bit more than a dozen amino acids. Yet the size of the scrapie protein implied that it had 250 amino acids.

Prion Proposal

Carleton Gajdusek was dividing his time between the NIH and his home in New Guinea, where, in 1978 and 1980, Prusiner made pilgrimages over rough terrain. "He arrived almost dead," Gajdusek told writer Richard Rhodes, "and stayed in my bush house for a couple of nights. We were in continuous bull sessions for all that time, discussing the future of kuru and CJD and scrapie work."[13] Gajdusek said that by that time he had already come to the conclusion that the scrapie agent was a protein. In his discussions with Prusiner, Gajdusek told Rhodes:

> I pointed out to him that I would give the disease agents a proper name when we were sure what their molecular structure was. I made this point repeatedly with him, explaining that it was premature to name them. . . . I had not realized that Stan would not give me the prerogative of naming them when the appropriate information was at hand. It was a clever political move on his part to jump the gun.[14]

Here, Gajdusek was referring to a conference presentation Prusiner made in February 1982 and the subsequent paper he published in the April 9, 1982, issue of *Science*, one of the premier journals in the world. "Novel Proteinaceous Infectious Particles Cause Scrapie," Prusiner declared in the title. In the paper, Prusiner summarized his attempts to purify and enrich scrapie samples and the outcomes of various chemical and physical assaults on the agent, all of which suggested that nucleic acids weren't present.[15]

> Because the dominant characteristics of the scrapie agent resemble those of a protein, an acronym is introduced to emphasize this feature. In place of such terms as "unconventional virus" or "unusual slow virus-like agent," the term "prion" (pronounced *pree-on*) is suggested.[16]

He argued that several hypotheses, such as viroids and replicating polysaccharides, were no longer viable, but he did not altogether come out in favor of the idea that the scrapie agent was only protein. He emphasized that a small nucleic acid could exist within the tightly packed protein coat. "Rigid categorization of the scrapie agent at this time would be premature," he wrote.

Others saw it differently. By introducing "prions," Prusiner was clearly promulgating a protein-only concept. The British journals *Nature* and *The Lancet* reacted indignantly, pointing out that Prusiner hadn't really explained anything new about the agent and that the introduction of his term only confused matters. Prusiner's lab colleague at the time, biochemist Frank Masiarz, refused co-authorship of the paper, feeling that there was no point in naming something before its existence was even known.[17] Later on, though, researchers in Prusiner's lab found more circumstantial evidence, showing that the more prion protein (dubbed PrP) there was in a scrapie tissue sample, the more infectious the sample was.

Fatal Filaments

While Prusiner made biochemical strides in isolating the scrapie agent, on the Atlantic side of the continent Patricia Merz of the New York State Institute for Basic Research on Staten Island, New York, was zeroing in on an important clue through microscopy. After teaching herself to image objects on the electron microscope, which involves complex staining and other preparations of samples to achieve good photographs, Merz decided to see if she could spot the scrapie agent, something that had eluded researchers because it was so difficult to purify samples. She collaborated with Robert Somerville of the U.K. Institute for Animal Health in Edinburgh, who provided her with tissue samples from scrapie-infected rodents. She began looking in February 1978 and soon spotted something unique to scrapie samples: tiny, stick-like filaments, each consisting of even tinier filaments. Most of the fibrils were a few tens of a nanometer (billionths of a meter) wide and ran a few hundred nanometers long, although some reached 1 micron (millionth of a meter) in length.[18] The more advanced the disease in the animal, the more it seemed to have these sticks. Later, in studies with the husband-and-wife team of Elias and Laura Manuelidis, medical researchers at Yale University, Merz found that these sticks also appeared in brain and spleen samples of CJD patients.

In her 1981 journal write-up with Robert Somerville, Henry M. Wisniewski, and Khalid Iqbal, Merz named the objects scrapie-associated fibrils, or SAF. No one knew what they were—Merz wondered if the fibrils were made of amyloid, the massive accumulation of which produced the plaques characteristic of many TSE diseases. But the scientists that Merz consulted didn't think so. Indeed, the sticks failed a

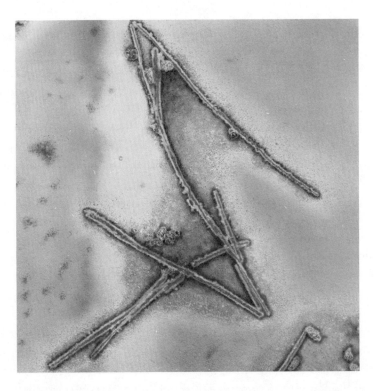

Scrapie-associated fibrils (SAF), also called prion rods, are revealed in this image taken through an electron microscope and magnified by 100,000 times. (*EM Unit, VLA/Photo Researchers, Inc.*)

crucial amyloid test. When amyloid is stained with the dye Congo red, it displays a sparkling called birefringence—it shifts from green to gold when viewed under different orientations of polarizing filters. The SAF failed this test, although Merz suspected that the failure had to do with the impurity of the sample.

Her suspicions proved to be correct. Two years later, Prusiner reported seeing prion rods that formed tubes 10 to 20 nanometers wide and 100 to 200 nanometers long. They looked remarkably the same as the SAF. With his better purifying techniques, Prusiner showed that these rods did indeed display the characteristic color shift of amyloid. Later, he found that they consisted of an extremely stable core of PrP molecules, perhaps up to 1000 of them stacked together like Lego blocks. (Prusiner has steadfastly denied that SAF and prion rods are synonymous, but most other researchers say otherwise.)[19]

The association of fibrils with scrapie-infected tissue didn't mean that the rods were the causative agent. Indeed, British researcher Harash K. Narang apparently spotted these rods well before Merz and Prusiner did but couldn't really characterize them or prove them to be the scrapie agent. In fact, the fibrils could be by-products of infection instead. Although the presence of the SAF correlated with the degree of infectivity of the sample, some samples proved to be infectious even though no fibrils were visible. Prusiner also found in 1991 that, in test tubes at least, prion rods formed as a consequence of purification and the presence of detergents, which evidently caused the PrP molecules to clump together to form the rods.

The Normal and the Diabolical

One step in proving that the protein was the infectious agent was identifying the sequence of amino acids that make up the protein molecule. Such protein sequencing involves a reagent that latches onto the amino acid at the end of the protein molecule, another compound to knock that amino acid off the molecule, and then a chemical to extract the amino acid for analysis. Chromatography, which separates chemicals as they filter through a material at different rates, helps to identify the amino acid. The process is repeated on the remaining protein molecule until the complete amino-acid sequence is determined. In sequencing PrP, Prusiner and his collaborators first found glycine, then glutamine, then a run of glycine, followed by threonine and histidine, and so on, until they had the first 15 amino acids of PrP.

Once you have the amino acid sequence, you can work backward to figure out the messenger RNA sequence that coded for those amino acids. And once you deduce the RNA sequence, then you can deduce the DNA sequence that coded the messenger RNA. The DNA sequence can be synthesized in the lab by joining nucleic acid segments referred to as oligonucleotides. Once synthesized, it can act as a molecular probe for natural DNA.

Prusiner enlisted the help of Leroy E. Hood of the California Institute of Technology and Charles Weissmann, then of the University of Zurich—both giants in the field of molecular biology—to create the DNA probes for PrP. (Automated machines can now make oligonucleotide probes.) To Prusiner's astonishment, the probe revealed that PrP existed not only in scrapie-infected hamster brains, but also in completely healthy tissue.[20] About the same time, Bruce Chesebro of the NIH Rocky Mountain Laboratories and his colleagues also made their own molecular probes and found that PrP existed in normal mice.[21] In fact, all species of lab animals examined harbored a gene for PrP—it even showed up in humans, on the short arm of chromosome 20 (the gene was later dubbed PRNP). The gene seemed to be active in most cells of the body, though it was particularly vigorous in some areas, including the brain and the heart.

Prusiner sat on the results for months. "One interpretation of such findings was that we had made a terrible mistake: PrP had nothing to do with prion diseases," Prusiner reasoned. "Another possibility was that PrP could be produced in two forms, one that generated disease and one that did not."[22] Prusiner and his colleagues soon showed the latter to be correct when they added a detergent called proteinase K, a commonly used, powerful enzyme that breaks apart most proteins. PrP from healthy tissue was destroyed, but that from scrapie tissue resisted the enzyme. Specifically, the PrP from scrapie tissue had a protease-resistant core of about 27,000 to 30,000 daltons. So even though they were chemically identical, consisting of the same amino acids, the PrP in normal cells was clearly different from the PrP in scrapie cells. That difference could only stem from the way the two forms were folded.

The normal prion protein must be folded in such a way that it could be dissolved by proteinase K; the pathological prion protein, on the other hand, must have adopted a conformation that resists the enzyme. Prusiner introduced the abbreviations PrP^C to denote the former (C for cellular) and PrP^{Sc} for the latter (Sc for scrapie, although it now refers to the pathological prion protein from any TSE). He used PrP 27-30 for the protease-resistant core.

The idea that the normal prion protein PrP^C could be reshaped into the diabolical version PrP^{Sc} went against the orthodox view of infectious disease. But it would explain why some spongiform encephalopathies are inherited. In fact, by 1989, researchers would show that the familial form of CJD results from an abnormal prion protein gene—and that two other, rarer hereditary TSEs also stem from mutations of the gene that passes down from one generation to the next. One of these illnesses generally produces ataxia and mental decline that may evolve to severe dementia before death ensues around five years later. The other results in the loss of the ability to sleep, leading to near madness from permanent insomnia, and death within a few months.

CHAPTER 6

Family Curses

Two rare hereditary diseases add support to the prion hypothesis—and challenge it, too.

She was a daughter in the "H" family, a bloodline that cursed each generation with a prion disease. Her great-grandfather had it, as did both her grandfather and her father. Her condition puzzled her Austrian physician Josef Gerstmann, who had studied neurology and psychiatry in Vienna, which—thanks to Sigmund Freud—was the place to be in the early twentieth century to learn about the troubled mind. And like Freud, Gerstmann saw the dangers of the emerging Third Reich, leading him to flee to the U.S., where he died in 1969.

In 1928, the 41-year-old Gerstmann described Miss H: "The disease syndrome of a 26-year-old patient, who has been under my observation at the Vienna Psychiatric Clinic for some time . . . is quite noteworthy," he wrote of the young woman. "The disease broke out in 1926 with an initially severe, over time, gradually worsening imbalance while standing and walking. The illness came as such a surprise that relatives, owing to the patient's insecure, staggering walk, initially thought she was drunk." Relatives soon noticed personality changes. "She became irritable, irascible, intolerant, furious" at times, Gerstmann wrote, but "her mood was usually cheerful for no reason."[1]

Clinically, the main symptoms were not dissimilar from those of Creutzfeldt-Jakob disease.

> The patient stands and walks with her legs apart, she deviates from a straight line while walking, she sways and staggers in various direc-

tions. . . . Besides there is an unmistakable decrease in intelligence of a progressive nature that has already advanced to a considerable degree of dementia.

Most vexing, Gerstmann found, was a reflex action when the patient's arms were extended in front of her. Turning her head—either by her own accord or by the force of the physician's hands—caused her arms to swing across her body. Both arms would be held out at one side, one above the other, as if she were trying to block someone from passing her. When Gerstmann turned her head to the right, her arms swung to the left, and vice versa.[2] She died at age 31, six years after the onset of symptoms.

This patient and seven other affected family members were autopsied, and in 1936, Gerstmann, along with fellow Austrians Ernst Sträussler and I. Scheinker, described the pathology of their diseased brains. Under the microscope, they saw plenty of neural degeneration and loss of cells. In the cerebral cortex, they spotted gliosis (proliferation of infection-fighting cells of the brain) and "Lücken"—holes in the tissue.[3]

In reviewing the case histories of several patients suffering from what is now generally referred to as Gerstmann-Sträussler-Scheinker syndrome (GSS) in 1962, Austrian neurologist Franz Seitelberger noted that it was primarily a motor disease that "starts in the fifth decade of life, in a few cases also in the fourth. . . . Patients become bedridden and die in marasmus or of intercurrent infections [they waste away or develop secondary infections] after a total disease duration of two to seven years."[4] Seitelberger remarked on the accumulation of amyloid plaques, noting "deposits affecting the cerebral cortex and the basal ganglia, also extremely affecting the cerebellar cortex in all its layers where they differ in fine structure from the typical shape of plaques. Clinically, these deposits in the gray matter can be related to psychological changes of presenile progressive dementia." GSS occurs at a rate of about 1 per 15 million people and is now known to affect at least four dozen families spread across nations—besides Austria, they include France, Britain, Japan, Germany, Sweden, the U.S., Canada, and Mexico.

In explaining the symptoms and features of GSS, Seitelberger connected various aspects of it to other neurological diseases. He zeroed in on one:

The most striking relationship, however, exists between the native here-doataxia [GSS] and a neurological disease called Kuru. . . . The Kuru findings are important for our study because they demonstrate once more the coupling of two different histopathological syndromes.

Seitelberger, however, was working from the assumption that kuru was a genetic disease. (Unlike William Hadlow, Seitelberger did not think it was worth pursuing kuru's connection to scrapie.) The kuru–GSS connection proved to be more remarkable than Seitelberger realized. In the 1960s, Gajdusek, Gibbs, and Colin Masters (a visiting researcher at the NIH) proved that CJD could be transmitted to several kinds of experimental animals. In 1973, the three scientists reported the stunning news that they successfully transmitted familial forms of CJD—and GSS in 1981—into monkeys.[5] Spongiform diseases could evidently be both infectious and hereditary.

Coding for Disease

The earliest explanation for the GSS transmission studies was straightforward: Perhaps GSS families inherited a genetic susceptibility to the slow virus that also caused kuru and sporadic CJD. That would be the most natural assumption—if viruses were in fact the cause of TSEs. But in the protein-only conception of Stanley Prusiner's prions, viruses have no place in the pathology.

The modern tools of molecular biology enabled Prusiner and his colleagues to determine the DNA sequence of the gene for the prion protein. All they had to do to prove his theory was to induce the prion gene to make its corresponding protein, PrP, and then show that the prion protein could cause disease. "By 1986, however, we knew the plan would not work," Prusiner recalled. "For one thing, it proved very difficult to induce the gene to make the high levels of PrP needed for conducting studies. For another thing, the protein that was produced was the normal, cellular form"—PrP^C, rather than the infectious, "scrapie" form, PrP^{Sc}. "Fortunately, work on a different problem led us to an alternative approach."[6]

That approach was to look at clearly inherited TSEs. In 1988, Prusiner and Karen Hsiao, one of his graduate students, obtained clones of the PrP gene from a man dying of GSS. They compared his gene with PrP genes obtained from healthy people and found a tiny abnormality in the patient's PrP gene. This abnormality, additional research showed, also existed in genes from several GSS patients, including the "H" kindred that Gerstmann first described. "We established genetic linkage between the mutation and the disease—a finding that strongly implies the mutation is the cause," Prusiner concluded.[7]

The genetic blip was a point mutation—that is, a single pair of bases in the prion protein gene differs from the pair found in healthy people. Base pairs—A and T, C and G—make up the "rungs" of the DNA double helix. It takes three base pairs to form a codon, which indicates a particular amino acid. (Virtually all living things rely on just 20 kinds of amino acids to make all the proteins they need.) Codons can also be thought of as tags that mark locations on the protein. For example, codon number 129 corresponds to where you would find the 129th amino acid of the protein—in the case of the prion protein, it could be the amino acids methionine or valine.

In the "H" family, one altered base pair out of the more than 750 base pairs produced a different amino acid at codon 102 (253 amino acids make up PrP^C). Family members had thymine swapped for cytosine, resulting in an altered code that produced the wrong amino acid. At the codon 102 position, healthy individuals have the amino acid proline, while the "H" and other GSS families had leucine. Leucine in the place of proline changes the properties of the prion protein, making it fold up differently and thereby behave like the pathological form, PrP^{Sc}. (Researchers abbreviate point mutations by using the two amino acids' letter designations around the codon number—hence, this mutation where proline gets substituted by leucine at codon 102 is written as P102L.) Soon after identifying this mutation, Hsiao added further evidence when she created a breed of mice that produced the mutant GSS form of PrP. These transgenic mice spontaneously developed spongiform disease.

The proline-to-leucine substitution at codon 102 affected not only the "H" family but also other Gerstmann-Sträussler-Scheinker syndrome kindred in the U.S., U.K., Germany, Italy, and Japan. But it's not the only mutation that leads to GSS. Researchers later found that other

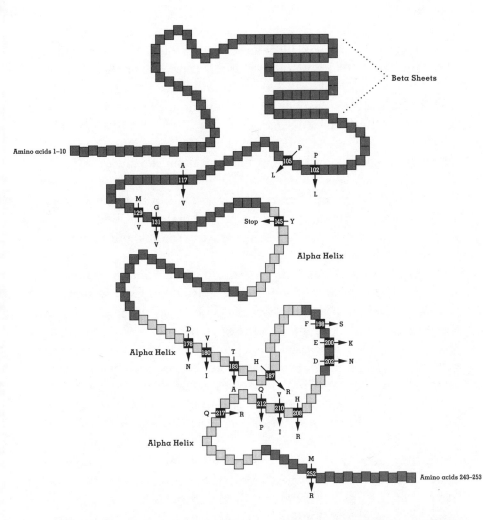

Human prion protein (PrP) is shown as blocks of its 253 amino acids. The amino acid substitutions that cause inherited prion diseases are labeled at their positons in the protein (or codons): the change from proline (P) to leucine (L) at the 102nd amino acid (codon 102), for instance, produces Gerstmann-Sträussler-Scheinker syndrome. The polymorphism at codon 129, where either methionine (M) or valine (V) can exist, influences the duration and symptoms. The mutation at codon 145, where tyrosine (Y) is substituted by a so-called STOP codon, result in a truncated protein. Disease can also result when PrP is made abnormally long, as can happen when mutations occur between codons 50 and 91 in the form of extra amino acids called octapeptide repeats that are inserted (not shown). Lightly shaded areas represent the part of PrP that coils up into alpha helices; the flattened areas are beta sheets. The other amino acids shown are: alanine (A), aspartic acid (D), phenylalanine (F), glycine (G), histidine (H), isoleucine (I), lysine (K), asparagine (N), glutamine (Q), arginine (R), serine (S), and threonine (T). (*After a concept by the World Federation of Scientists.*)

amino acid substitutions, such as an alanine-to-valine change at codon 117 (A117V), exist in GSS bloodlines. Families passing on the inherited form of Creutzfeldt-Jakob disease, accounting for 10 to 15 percent of CJD cases, were also passing on point mutations, such as the replacement of aspartic acid with asparagine on codon 178 (D178N) or glutamic acid with lysine at codon 200 (E200K). Such genetic analyses helped to eliminate some of the speculation about the cause of CJD. In 1974, for instance, researchers postulated that the consumption of undercooked sheep brains and eyeballs was responsible for the high incidence of CJD among Israeli Jews of Libyan descent. But it turned out that this population has a high incidence of the glutamic acid-lysine substitution at codon 200.

By the late 1990s, scientists discovered 13 point mutations on the prion protein gene that produce disease. Nine other types of mutations also produce disease. One is a change at codon 145: Instead of tyrosine, there is a STOP codon, which marks the end point of protein construction. The result is a truncated prion protein, 60 percent shorter than normal. The other eight mutations create an unnaturally long prion protein. They involve adding extra, so-called octapeptide repeats: the insertion of 24 base pairs in as many as nine additional groups between codons 51 and 91. (The normal prion gene has five octapeptide-repeated groups.)

The different mutations affect the clinical course and neuropathological picture in different ways. Some strike victims down in their 30s; others don't start until carriers reach their 60s. Some lead to disease durations of a matter of months; others go on for years, sometimes a decade or two. Some mutations send patients into dementia; others just produce ataxia and slurring. Some leave the characteristic plaques and holes in the brain; others produce few plaques and no holes. And one prevents its victims from sleeping, killing them after months of permanent insomnia.

The Family That Couldn't Sleep

A tall man with a slight hunch, Pierluigi Gambetti is quick with a smile and spurts of unexpected humor—at one meeting, he pointed to a colleague's expanded waistline and joked that he didn't need to come to lunch with us. But as we walked down a driveway outside his building on the campus of Case Western Reserve University in Cleveland, Ohio, in the fall of 2001, he looked straight ahead and furrowed his brow. In an Italian accent that hasn't leavened despite nearly 40 years in the U.S., he confessed softly, "I didn't think about it"—about the effect on patients when told the name of their affliction: fatal familial insomnia.[8]

It's a grim disease, one in which physicians can only watch helplessly as it singles out the part of the brain evidently involved in sleep. Intravenously administered barbiturates, able to induce sleep in seconds by blocking certain neural receptors and potent enough for general anesthesia, have little effect. "They succeeded in one case to get him to sleep," Gambetti said of one patient. "For half an hour."

Gambetti had signed off on the name of the disease as part of a collaboration that introduced the illness to the medical world in a 1986 case report in the *New England Journal of Medicine*. The patient, Silvano S., was an industrial manager who hailed from a small town outside Venice. Handsome and broad-shouldered, he began to notice some unusual symptoms shortly after a vacation with his mother, when he was 52 years old. Silvano normally slept seven to nine hours a night and napped 30 minutes in the afternoon; now, he was getting no more than two or three hours nightly. The acknowledged "ladies man" lost his sex drive and became impotent. Silvano's condition worried his relatives. They had seen this before—it wouldn't be long before Silvano would lose all ability to sleep.

At least 30 other members of his family line had gone down that torturous route. The first recorded death from insomnia was Silvano's great-great-grandfather Giacomo. Born in 1791, he begat several generations of descendants who would rise in Italy's socioeconomic ranks to prominent positions in medicine, business, and real estate. The family also developed an outcast reputation among the locals, as D. T. Max described in a *New York Times Magazine* article in May 2001:

> Within the Veneto region of Italy, where most of the family still lives, the knowledge of a family cursed with a strange disease has long been widespread. Villagers speak of it behind the family's back. Although the women tend to be beautiful and the family cultured and wealthy, finding spouses is difficult. The family cannot get life insurance.[9]

Silvano's niece, Elisabetta Roiter, told Max that she tried to buy insurance, "and after filling out the form, the woman in the office asked, 'So, at what stage are you in the family disease?'" The illness follows an autosomal dominant pattern: *autosomal,* meaning that the defective gene rode on a non-sex chromosome (neither X nor Y); and *dominant,* meaning that that the effects of the defective gene cannot be masked by a healthy gene (as is the case for recessive genetic disorders, such as cystic fibrosis). A child of parents who each have the defective gene would be assured of getting the defective gene; if just one parent has the defective gene, then it's 50–50 whether a child will get it. So when Uncle Silvano began showing symptoms of the family curse, Elisabetta was devastated. She shared one-fourth of his genes, so she had a 25 percent chance of getting it herself. If her mother, Silvano's sister, had the disease, her odds would be 50–50. She and her husband Ignazio wanted to have children but didn't want to pass on the feared affliction. Elisabetta would often sneak into her mother's bedroom at night to make sure she was really asleep. "She got annoyed and started throwing her slippers at me," Elisabetta recalled.[10]

Trained as a nurse, Elisabetta was ideally suited to get to the bottom of the family curse. Her grandfather Pietro, the mayor of his hometown in Veneto under Mussolini, died in 1944 shortly after receiving a death threat after the Fascist government collapsed. The cause was listed as encephalitis. "The family just accepted these judgments," Elisabetta said. "And we had our own myths. My grandmother, for instance, called it 'a disease of exhaustion,' because she believed it struck you after a moment of extreme stress."[11] Looking at Pietro's chart during a 1971 visit to the hospital where he died, she saw an odd notation. It said that Pietro's cerebrospinal fluid was clear—but encephalitis tends to produce a cloudy fluid. A few years later, one of Elisabetta's aunts complained of depression and insomnia; a neurologist diagnosed her with dementia, even though she was able to understand everything around her. Another relative supposedly died of schizophrenia.

Two months after he first complained of symptoms in 1984, Silvano managed only one hour of sleep a night. His dreams became more vivid, and he began acting them out. Once, he got out of bed, stood, and saluted, believing he was at a coronation. A month later, sleep became impossible. Fatigue dominated his waking days, and he began to slur his words. Six months after his symptoms started, Silvano had difficulty breathing and walked clumsily.

Desperate, Elisabetta and Ignazio, who was studying to become a physician, contacted Elio Lugaresi, who ran the Neurological Hospital of the University of Bologna Medical School. Lugaresi admitted Silvano the next day, giving him a comfortable bed, setting up a video camera, and wiring his brain with electrodes. The videotape, Max wrote, "makes for uncomfortable viewing. His course is relentlessly downward. On a tape made in March [1984], his eyelids flutter over the dots of his eyes."[12] (Silvano's pupils had contracted and reacted only weakly to light.) As described in the *New England Journal of Medicine* case report, Silvano exhibited "brisk deep-tendon reflexes": whack the spot just below his knee with a small hammer, and his leg would unleash a kick that was unusually forceful. Lugaresi could also feel some fine trembling in Silvano's arms. Both are signs that his motor neurons were deteriorating. If Silvano was left alone, he would lapse into a stuporous, dreamlike state, sometimes gesturing boldly as if acting out whatever it was his brain was fantasizing.

One month later, when Silvano was admitted for the second time, the neurological signs had become even more obvious: amnesia, twitching of the limbs, an inability to maintain a straight gaze. His sleep-deprived body shifted into overdrive: He ran a temperature of 100.4° F and his heart throbbed at 100 beats per minute while he was sitting still. By the eighth month after the symptoms began, he was mostly stuporous, but he had brief episodes in which he howled and jerked, his muscles contracting involuntarily. His hormone levels were all over the place. Normally, hormone levels wax and wane through the day and night to regulate the body's functions. But Silvano's endocrine system refused to keep pace with the circadian rhythms and so released cortisol, growth hormone, melatonin, and other endocrine products at a constant rate. Silvano's body was out of control, and it wasn't surprising that a pulmonary infection set in. Nine months after symptoms began, Silvano lapsed into a coma and died, his sleepless descent finally ending.

Lugaresi had Silvano's brain removed, preserved in formaldehyde, and shipped to Gambetti, who had done his medical residency at Lugaresi's hospital in the early 1960s before coming to the U.S. (he has been directing the neuropathology lab at Case Western since

1977). After sectioning the brain into hundreds of thin slices for microscopic examination, Gambetti found that the damage seemed to be confined to two areas of the thalamus (the anterior and dorsomedial nuclei of the thalamus). There was a substantial loss of neurons there—about 95 percent of them were gone—and a two- to threefold increase in astrocytes. Other parts of the brain, including the cerebrum, were normal.

Silvano's case report, which also included some details about a sister who died in similar fashion in 1978, represented the first attempt to characterize the disease, and it left Lugaresi, Gambetti, and their colleagues puzzled. The pathological changes did not correspond to those seen in any form of Creutzfeldt-Jakob disease. Nor did the findings mesh with previously described brain atrophies that involve the thalamus. The data also intrigued sleep researchers. No one knows exactly how the brain triggers slumber—an area of the hypothalamus called the suprachiasmatic nucleus plays an important role as a sleep trigger and as a metronome for the body's circadian rhythms. Scientists didn't think that the thalamus was involved. Thalamus means "antechamber"—it's a gateway between the brain stem and various parts of the cerebral cortex. Its primary duties are to relay sensory signals and help in memory formation along with other parts of the brain. Silvano's brain suggested that the anterior and dorsomedial parts of the thalamus are vital for sleep and for normal endocrine and circadian functions.

More data enable researchers to improve their suppositions. In the case of fatal familial insomnia, or FFI, that meant obtaining more brains. In 1992, Lugaresi, Gambetti, and their colleagues reported five new cases from the family. Teresa, a 35-year-old distant cousin of Elisabetta's, was the youngest to come down with the disease. A mother of two, she suffered sleeplessness, hallucinations, dream enactments, and twitching over the course of 25 months before she died. Her brain disintegrated in the same areas of the thalamus as Silvano's. What was different, however, was her cerebral cortex: It showed spongiform change, ranging from a delicate separation of tissue in some areas to full-blown holes in others. Evidently, the lengthy course of Teresa's illness made spongiosis apparent. "The question must be raised," the team stated in their write-up of the cases, "as to whether FFI is a prion disease with a pathologic phenotype similar to that of CJD."[13]

The question was asked, and genetic analysis provided the answer. For the analysis, the scientists relied on the brains of Teresa and an afflicted cousin, which had been frozen after death, and blood samples from 33 other family members. They found that those with the disease shared a common point mutation, one riding on the prion gene— specifically, at codon 178. Normally, the base sequence there reads guanine-adenine-cytosine, specifying the amino acid aspartic acid. But in FFI patients, the codon reads adenine-adenine-cytosine, specifying asparagine. A single base pair out of the 3 million base pairs in human DNA was enough to produce an incurable insomnia.

Knowing what to look for, the investigators could now determine who might develop FFI and who would be spared. Elisabetta turned out to be negative. (She had guessed as much—by the time Silvano died, her mother turned 65, beyond the age at which FFI usually strikes. Elisabetta and Ignazio had a daughter in 1986.) Ultimately, tests for half of 50 relatives came back positive for the mutation.

Genetic analysis also made it apparent that fatal insomnia wasn't restricted to Giacomo's descendants. Throughout the 1990s, researchers found that the mutation occurs around the world, affecting at least 27 families. Most are found in Europe, especially Italy, France, Germany, Austria, and England, and a few in Japan. Australia and North America, too, have FFI families of German and Chinese extraction. Fatal insomnia also occurs sporadically: 7 cases have emerged in which there was no family history or PrP mutation. Overall, fatal insomnia is an exceedingly rare disease, showing an incidence of about 1 per 33 million people.

One Codon, Two Diseases

What was curious about the mutation at codon 178 was that some people who had the mutation came down with a different disease. They developed classic CJD instead of FFI. How could the same mutation produce two diseases with distinctly different clinical courses and brain damage? It didn't take long before researchers discovered the answer:

Along with the codon 178 mutation, there was another change at a different location in the prion protein, at codon 129.

Codon 129 turns out to mark a rather special position in the human prion gene. A polymorphism can exist here—that is, two codings are possible, one calling for methionine (M) and an alternative specifying valine (V). Either amino acid yields a fully functioning prion protein, all else being normal. Since genes occur in pairs called alleles, with one allele coming from each parent, an individual may possess one or both amino acids at codon 129. If both your alleles code for methionine, or both code for valine, then you are homozygous at codon 129 (abbreviated as M/M or V/V). If one allele codes for methionine and the other for valine, then you're heterozygous (M/V).

Among Caucasians, the methionine/valine combination dominates the population: 51 percent are M/V. About 37 percent are M/M and 12 percent V/V. In Japan, however, the valine allele is so rare, it's essentially a mutation. In fact, one study found no Japanese subjects who were valine homozygous. Rather, 92 percent are M/M and 8 percent are M/V.

The three possible variations at codon 129 determine whether a patient with the codon 178 mutation develops FFI or CJD. A patient with valine at codon 129 develops ataxia, myoclonus, and other typical signs of CJD. But if methionine occurs at codon 129, from either an M/M or M/V combination, then fatal insomnia develops. Codon 129 also governs the duration of FFI. People with the M/M combination, like Silvano, die within about a year, and disturbances of sleep and autonomic functions are pronounced. People with the M/V combination, like Elisabetta's cousin Teresa, last twice as long; motor problems are more apparent, sleep disturbances less so.

In hereditary illnesses, researchers often refer to *penetrance*—the odds that the mutation will actually produce disease. Single-gene diseases such as Huntington's disease are 100 percent penetrant: if you have the mutation, you will develop the neurological disease. Inherited diseases that ride in on several defective genes acting in concert may have low penetrance. For prion diseases, the penetrance factor appears to be quite high. For FFI, it is above 90 percent, Gambetti said. The codon 200 mutation of CJD has been cited as 90 to 100 percent. It's not always easy to determine penetrance, because not all family members permit themselves to be genetically tested. But in all likelihood, if you have a prion gene mutation and you live long enough, you will get a prion disease.

The Strains Puzzle

The mutations in the prion protein that lead to GSS, FFI, and familial CJD buttressed the foundation of Stanley Prusiner's theory that a protein could transmit disease. A single miscoding in the DNA resulted in a prion protein that folded up differently, and this misshapen protein, if inoculated into a healthy individual, could cause that individual's prion protein to adopt a malformed shape.

Other support for the prion theory came from studies of genetically engineered lab animals. With the modern tools of molecular biology, scientists are adept at manipulating the genes of mice, flies, worms, and other organisms. They can selectively remove DNA, effectively "knocking out" a target gene. Alternatively, they can introduce an alien gene that blocks the expression of the target gene. Researchers can also substitute genes—for example, scientists can remove a mouse gene and replace it with the corresponding human version. In lab experiments, such "humanized" mice better mimic what might be going on in people. The genetic manipulations are done on embryos, so that these transgenic mice, once fully grown, can pass on the foreign traits to their progeny.

Transgenic mice have generated a good deal of evidence in support of the prion theory. Those that were genetically engineered to express human prion protein got infected when given brain homogenates of people who died of an inherited spongiform encephalopathy. What's more, transgenic animals that expressed the mutant form of the human prion protein became sick on their own.

The converse is true as well: Mice whose natural prion gene was knocked out did not develop disease when inoculated, in contrast to their normal, "wild-type" cousins. According to the protein-only theory, the knock-out mice could not be infected because their bodies did not produce any prion protein, and so there was nothing for the introduced agent to convert.

Despite being persuasive, the evidence from human TSEs and animal studies didn't preclude the possibility that a virus or other nucleic-acid-bearing entity was playing a role. Skeptics of the prion theory correctly pointed out that the mutations may simply have made the carriers susceptible to a prevalent virus. Conceivably, a mutant prion protein simply provided the gateway through which a virus or other foreign piece of nucleic acid could enter and wreak havoc on the central nervous system. So, one way to explain the results of the "knockout" mouse studies was that the lack of prion protein simply meant that the virus did not have a suitable protein receptor to latch onto, and thus it had no way of infecting the animals.

Basically, the issue boiled down to strains of the disease. The existence of viral and bacterial strains is the reason why annual flu shots are necessary, and why some *E. coli* are deadly and some benign. The slight variations in the genome are too small to turn the pathogen into a different species, but they are sufficient to alter some of its characteristics.

Almost as soon as Prusiner announced his prion hypothesis, critics pointed out that the protein-only concept couldn't explain the presence of strains of transmissible spongiform encephalopathies. In the lab, researchers documented some 20 TSE strains, based on incubation times and patterns of brain lesions in rodents. GSS, FFI, and CJD produce different symptoms, incubation times, and patterns of destruction in the brain. CJD itself has several variations—for instance, in the Heidenhain strain of CJD, patients go blind because of degeneration of the brain's occipital lobes, which process visual signals from the optic nerve.

If the prion protein came in a good form and a bad form, how did the bad form create so many different variations? Only nucleic acids could encode strain information, geneticist Alan Dickinson of the Neuropathogenesis Unit in Edinburgh pointed out in a *Lancet* editorial that appeared shortly after Prusiner introduced the term "prion" in 1982. Proteins had no known mechanism by which to encode the information needed to produce various strains. If the TSE agent was just abnormal prion protein, argued Moira Bruce, also of the Neuropathogenesis Unit, in a BBC TWO interview, "then the existence of different strains and the requirement for an informational component that specifies strains means that the protein itself has to carry that information. It's very difficult to envisage how it could do that."[14] Prusiner postulated that PrP^{Sc} could fold up in slightly different ways, thereby adopting many shapes, or be modified by the addition of other molecules.[15] "But this would involve a very faithful reproduction of that shape over very many cycles of reproduction and in that case there would have to be as many conformations as there are different strains," Bruce argued.[16]

One counterpoint is that there aren't really that many strains to begin with. The husband-and-wife team of Rosalind Ridley and Harry

Baker, veteran TSE researchers at the University of Cambridge, traced the origins of strains as researchers transmitted the TSE agent through different animals. In their review, they concluded that many strains derived from the same pool of scrapie brains (termed SSBP/1). Transmitting SSBP/1 to mice resulted in strain line "22"; putting it into goats resulted in "drowsy" and "scratching" strains (named after the predominant clinical sign). Transmitting "drowsy" into mice produced the Chandler strain, which on further mouse transmissions yielded mouse strains 79A and 87V. Using the "scratching" strain in mice produced the "22" strain again.

One explanation for the results is that the "drowsy" source was contaminated with the other strains, so that transmitting the source into mice simply revealed the hidden strains. Or perhaps mutations of a purported virus occurred, thereby creating new strains. In the prion hypothesis, however, the reason for the diversity is that each host— sheep, goat, mice—have different prion proteins, as to be expected of different species.[17] The malformed prion protein, PrP^{Sc}, doesn't have to encode strain information—it would "get it" from the infected host. The introduced PrP^{Sc} would cause the host's own prion protein to misfold in a species-specific way. That would at least explain how the same brain material could affect each species differently. But that wouldn't explain why the clinical picture varied in the same species.

In the early 1980s, Dickinson, along with colleagues George W. Outram and Richard Kimberlin, proposed an alternative theory that fit the data—a mini-virus that lacks its own protein coat. Entities called viroids fit the bill: They are bits of RNA that, unlike viruses, do not have a coat of protein. They are about a tenth the size of the smallest known viruses and infect only plants. But the three scientists were thinking of something even smaller than viroids—it would *have* to be, because the scrapie agent still hadn't been seen. Drawing an analogy to the way physicist Enrico Fermi named a nearly massless particle similar to the neutron ("neutrino," for little neutron), Dickinson termed his hypothetical mini-virus a virino. As a bit of noncoding nucleic acid, the virino wouldn't produce protein on its own; instead, it would rely on its host for its protein coat and thereby escape detection by a healthy immune system, which targets foreign proteins only. How quickly the virino produced disease would depend on the host's genetic makeup; Dickinson found a gene in mice he termed *Sinc*, for *s*crapie *inc*ubation. (This gene was later shown to be the mouse prion gene.)

Explaining Strains with Prions

Considering that for two decades prion skeptics pointed to strains as the major gap of the protein-only hypothesis, I was surprised to hear from several researchers that the strain issue didn't bother them. That was true even among those not fully aligned with the protein-only hypothesis. According to the NIH's Paul Brown, Carleton Gajdusek filled the hole even before Prusiner coined the term "prion." "The idea that Carleton had many years ago," Brown explained, "was that the process was akin to crystal formation. He didn't say it *was* crystal formation, he said it was akin to it. And as we know, crystals can take any number of shapes depending on what circumstances the constituents find themselves in. I think that's a perfectly plausible hypothesis."[18]

The prion protein is a glycoprotein—a protein with sugar molecules attached to it. The two sugar groups attached to the protein are quite complex and account for one-third of the total molecular weight of PrP. In principle, more than 400 forms of PrP are possible based on changes in the sugar groups. "The notion that there would be different strains wedded necessarily to the presence of nucleic acid," Brown remarked, "is probably not necessary."

One of the first pieces of evidence that PrP^{Sc} had different versions that were responsible for strains was in early 1992, when Lugaresi, Gambetti, and their colleagues traced fatal familial insomnia to codon 178 of the prion gene. In dissolving the PrP^{Sc} from FFI and CJD patients with proteinase K, they found that the indigestible core of PrP^{Sc} came in two sizes. That from FFI left a "residue" of around 19,000 daltons, CJD of about 21,000. It implied that the two diseases resulted from different forms of PrP^{Sc}. Similarly, in December 1994, NIH Rocky Mountain Laboratories investigator Richard A. Bessen and Richard F. Marsh of the University of Wisconsin at Madison compared the PrP^{Sc} from hamsters infected by the "hyper" and "drowsy" strains of a TSE in mink (called, appropriately enough, transmissible mink encephalopathy). They found that proteinase K degraded hyper

PrP^{Sc} and drowsy PrP^{Sc} at different rates and cleaved them at different places. In a follow-up study, Bessen, Byron Caughey, and their colleagues found that the hyper and drowsy strains of PrP^{Sc} could convert PrP^C (normal prion protein) to those strains, at least in the test tube. "These data provide evidence that self-propagation of PrP^{Sc} polymers with distinct three-dimensional structures could be the molecular basis of scrapie strains," they wrote in *Nature*.[19]

The propagation of strain characteristics occurred in a cell-free system—but did it work in the morass of chemicals in the body? Compelling evidence emerged in December 1996, when Glenn C. Telling, Prusiner, Gambetti, and their colleagues reported results of experiments in which brain homogenates from FFI and CJD victims were injected into mice genetically engineered to express human prion proteins. Mice inoculated with FFI extracts developed 19,000-dalton PrP^{Sc}, whereas those that were inoculated with CJD extracts generated 21,000-dalton PrP^{Sc}. These molecular weights matched those from human victims of the two diseases. Moreover, the pattern of damage to the brains depended on the type of inoculant. Taken together, the experimental evidence strongly indicated that nucleic acid wasn't necessary for the existence of strains. Still, it wasn't proof that proteins alone could pass on inherited characteristics. An unseen virino could be hitching a ride with the various PrP forms.

The most persuasive evidence that proteins can possess inheritance information, passing on their shapes one generation after another, came from an unexpected source. Researchers in 1965 and 1971 had found mysterious hereditary patterns that standard genetic theory could not explain. The species showing these patterns and becoming an object of intense investigation is called *Saccharomyces cerevisiae*—better known as baker's yeast.

CHAPTER 7

On the Prion
Proving Grounds

*Research in yeast and other studies show how prions can
possess hereditary information and change their shapes.*

Writing in a 1994 issue of *Science*, molecular biologist Charles Weissmann commented that phenomena discovered in complex organisms,
such as mammals, acquire additional respectability if they are also
found in yeasts.[1] Like animal cells, yeasts have a nucleus within a contained cytoplasm. But being single-celled organisms, yeasts are much
simpler than animals and reproduce quickly, making them more
straightforward to study in the lab than animal cells. Manipulating the
DNA of yeasts tends to be easier, and tracking and identifying genetic
changes can be done readily. Depending on the culture medium on
which it is feeding, an engineered strain of yeast, for instance, may
change from white to red. The laboratory experiments with yeast cells
provided unquestionable proof that proteins, and not just genes, can
act as elements of inheritance.

Two particular yeast genes were responsible for the confirmation.
One, called [PSI] and discovered by British geneticist Brian Cox in
1965, tells the cell when to stop translating messenger RNA, thus controlling the yeast's ability to make protein. The other, found by French
geneticist François Lacroute in 1971, is called [URE3] and plays a role in
metabolizing nitrogen. (Discoverers of yeast genes get to name them;
by convention, it is a three-letter designation, generally related to the

gene's function, and usually followed by a number indicating its discovery order within that family of genes.) The brackets around their names are used to denote their unusual behavior, which confounded scientists for decades. Specifically, [PSI] and [URE3] do not get passed on the way other genes are. That is, the yeast genes failed to conform to conventional Mendelian genetics.

Gregor Mendel (1822–1884), the nineteenth-century botanist and Augustinian monk from what is now the Czech Republic, is often referred to as the father of genetics. His classic experiments with pea plants and how traits of the plant's seeds, such as being wrinkled or smooth, were passed on laid the foundation for modern genetic analysis. In particular, Mendelian genetics indicate the probabilities for the dominant and recessive traits that emerge in offspring.

To see how Mendelian genetics works, consider cystic fibrosis. This inherited disease causes the body to produce too much mucus, which can lead to life-threatening clogging of ducts and passageways in organs such as the lungs, liver, and pancreas. Cystic fibrosis is an autosomal disease, meaning that the defective gene rides on a non-sex chromosome (specifically, it's located on chromosome 7). A person might carry the defective gene but remain healthy. That's because virtually all human cells carry two copies of the genome—one from each parent. (The exceptions are sperm and egg cells, which carry one copy, and red blood cells, which don't carry any.) The healthy gene from one parent can mask the expression of the defective gene received from the other parent. The child, however, remains a carrier of the condition. Hence, cystic fibrosis is called a recessive genetic disease. To get the disease, a person must have two copies of the bad gene.

Mendelian genetics predict the odds of the disease being passed down. Say a carrier (who has one defective gene and one normal gene) marries a non-carrier (normal gene/normal gene). For their children, four different combinations are possible: two of normal gene/normal gene, and two of defective gene/normal gene. None of the children will develop cystic fibrosis, but odds are that one out of two will be a carrier for the disease. Now say two carriers have children (defective gene/normal gene crossed with defective gene/normal gene). Then the offspring combinations are defective gene/defective gene, normal gene/normal gene, and two of the defective gene/normal gene. One in four will develop the disease, and one in two will be a carrier. This pattern is typical of recessive genetic diseases.

The odds shift in dominant genetic illnesses, such as the prion diseases fatal familial insomnia and Gerstmann-Sträussler-Scheinker syndrome. Here, a healthy gene cannot mask the expression of the defective gene. A person with a defective copy and a healthy copy will come down with the disease; there are no "silent carriers" for it. If this person had children with a healthy individual, then the chances that a child will also inherit the defective gene and come down with the disease are 50–50.

The Mendelian inheritance rules are not restricted to diseases. They apply broadly to physical traits, such as hair texture or eye color. Most features, however, are influenced by several genes, complicating the odds calculations as well as adding variety to descendants' appearances. An organism's genetic makeup is called its genotype; its appearance is called its phenotype.

Mendelian patterns of inheritance also hold for the genes of other organisms, including yeast—except for two unique genes: [PSI] and [URE3]. They don't follow the conventional patterns of inheritance. Sometimes, the genes seem to disappear in the next generation, only to reappear later. More curiously, this phenomenon depends on the external environment. It is as if blue-eyed parents beget blue-eyed children in Sweden, but, after the children relocate to Brazil and marry other blue-eyed Swedes, they produce children who are brown-eyed. The yeast genetics, though, are even more astonishing than the human analogy because the changes happen in clones of the parent yeast cell. Yeasts can clone themselves by budding: They squeeze out a bit of cytoplasm and then part of the nucleus to create daughter cells. The daughter cells have the same DNA as the mother cell—yet may not have the [PSI] or [URE3] traits even if the mother cell has them.

Prions of Yeast

The puzzle of these two yeast genes languished unexplained for two decades. According to National Institutes of Health biochemist Byron Caughey, "Nobody cared, but it was a great puzzle nonetheless."[2] Various

explanations were offered—maybe the traits were encoded by mitochondria or rings of naked DNA called plasmids, or perhaps undiscov-
ered nucleic acids were lurking about within the chromosomes. None of
the speculations panned out. "All of a sudden, this light bulb went off in
Wickner's brain, and he wrote this wonderful explanation that blew
everything wide open," Caughey remarked of Reed B. Wickner's April
1994 *Science* paper.[3] The chief of the Laboratory of Biochemistry and
Genetics at the NIH's National Institute of Diabetes and Digestive and
Kidney Diseases, Wickner noted that the unusual behavior of [URE3]
could be easily explained by the prion concept.

Specifically, Wickner drew a connection between the protein produced by the [URE3] "gene" and an altered form of another yeast protein produced by the gene URE2. The protein of the normal URE2
gene, referred to as Ure2p, controls certain metabolic pathways that
determine the kinds of food that the yeast can eat. A mutation of the
URE2 gene causes the resulting protein to fold up slightly differently,
yielding an altered protein (call it Ure2p-mutant); this altered protein
changes the yeast's metabolism so that the yeast can process more
diverse foods—in particular, Ure2p-mutant enables the yeast cell to
feed on a nitrogen-containing chemical called ureidosuccinate. And as
Lacroute had noticed three decades ago, yeast expressing [URE3] proteins could also feed on ureidosuccinate, even though the yeast had the
normal URE2 gene (which should have prevented that feeding ability).
An explanation, Wickner had realized, was that the [URE3] phenotype
was not produced by a gene at all, but by a protein; moreover, the
[URE3] protein was exactly the same as Ure2p-mutant. The idea would
not only account for the feeding on ureidosuccinate, but it would also
explain why Ure2p-mutant and [URE3] never coexisted in the same
yeast cell.

With modern molecular biology tools, Wickner wrote in his 1994
Science paper, he was able to confirm the either/or nature of Ure2p-
mutant and [URE3] proteins and the identical behavior of yeast expressing Ure2p-mutant or [URE3] proteins. He also noticed that
[URE3] arose more frequently in yeast cells when a lot of the normal
Ure2p was around—there was more protein to be converted from
Ure2p into [URE3]. In this way, [URE3] could replicate itself, at the
expense of Ure2p. Moreover, Wickner conducted experiments in which
he "cured" [URE3] yeast: adding the chemical guanidine hydrochloride
disrupted the [URE3] protein, thereby making the yeast behave like its

normal self, in which it could not metabolize ureidosuccinate. Wickner concluded that the compound prevented Ure2p from refolding itself into [URE3] (identical to Ure2p-mutant).

What's more, the cure was reversible in the sense that [URE3] could spontaneously return to a yeast strain that was cured by the guanidine treatment. Such a reversible cure does not occur in conventional infections. For instance, if you were cured of a cold virus, you would have to encounter yet another cold virus before you got sick again. In contrast, yeast "cured" of [URE3] got "sick" again with [URE3] even though nothing was introduced.

Wickner's experiments showed that the [URE3] trait of yeast behaved exactly the same way as prions do in humans. We have a prion gene that codes for a normal protein, PrP^C, that can be refolded into the pathological protein PrP^{Sc}. Yeasts have a URE2 gene that codes for a normal protein, Ure2p, that can be refolded into [URE3]. A mutation in the PrP gene of humans causes cells to manufacture abnormally folded prion proteins (PrP^{Sc}); likewise a mutation in the URE2 gene of yeast results in Ure2p-mutant proteins (identical to [URE3]).

With his observations, Wickner summed up the three main criteria necessary to distinguish a prion from a nucleic acid: reverse curability, increased numbers of the prion if the normal protein is overproduced, and an indistinguishability between the prion and the protein produced by a genetic mutation.

These criteria also held true for the yeast prion [PSI], which represents a refolded version of a protein called Sup35p, made from the gene SUP35. While working at the Howard Hughes Medical Institute at the University of Chicago, Susan Lindquist and her colleagues carried out crucial experiments that elucidated the biochemistry of yeast prions. (She now directs the MIT Whitehead Institute for Biomedical Research.) For instance, in work published in January 2000, Lindquist and colleague Liming Li were able to create artificial prions. They took a portion of Sup35p and, using now-standard molecular biology techniques, fused it with a protein from a rat. "We chose the rat protein because it was quite different from anything found in yeast," Lindquist said. "And we showed that this protein that was completely foreign to yeast could, in effect, be turned into a new type of yeast genetic element"[4] — namely, a new prion. The yeast-rat chimeric prion behaved exactly like other prions, thereby confirming that protein alone can pass on heritable traits without any DNA or RNA being involved.[5]

Unlike mammalian prions, yeast prions don't cause disease. In fact, they could be an evolutionary adaptation. Whether or not a yeast cell expresses Ure2p protein or the alternate form [URE3] depends on the environment in which the organism finds itself. [URE3], for instance, may help yeasts survive when ammonia levels are high. Having traits encoded by prions also enables yeasts to switch quickly between one form and the other, which isn't possible for traits encoded by DNA. The latest studies indicate that there may be 20 kinds of yeast prions. If true, this finding suggests yeasts can quickly adapt to many conditions. (At least, conditions in the laboratory: the species of yeast studied, *Saccharomyces cerevisiae,* is not commonly found in natural environments, so no one knows whether prions truly confer a genuine evolutionary advantage.)

Although scientists refer to [PSI] and [URE3] as prions, they are a world apart from the PrP of animals. "They're really completely different entities from a biological point of view. And the proteins have nothing in common from a sequence point of view," explained Caughey, who works primarily with mammalian PrP. "What they have served as are proofs of principle"—that proteins can act as elements of inheritance.

From Helix to Sheet

Although yeast prions are not disease-related, they do act in ways that are similar to the prion-disease process in people and thereby aid in the study of prion diseases. In 1997, for example, Lindquist and her team found that although Sup35p is soluble, its alter ego [PSI] can string together into long, insoluble fibers similar to the plaques of prion disease patients.[6] These fibers also readily took up the dye Congo red, just as kuru plaques do. Moreover, Lindquist and her colleagues manipulated the SUP35 gene so that it mimicked a class of mutant human prion genes. They inserted repeated sections into SUP35, similar to some mutations of the human PrP gene (the insertions of additional, so-called octapeptide repeats, which end up making PrP too long). Lindquist found that the repeated sections in the gene cause the Sup35p proteins that were produced to fold into the [PSI] form—the equivalent of PrP^C turning into PrP^{Sc}. That finding matches the fact that some kinds of familial Creutzfeldt-Jakob disease and Gerstmann-Sträussler-Scheinker syndrome are triggered by insertions of additional octapeptide repeats.

Despite insights from yeast experiments and improved technology to probe molecules, scientists still aren't completely sure of the details involved when the normal prion protein changes to its diabolical version. It's not yet technologically possible to use a high-powered instrument to capture images of the molecular shifting from PrP^C to PrP^{Sc}. Instead, researchers rely on indirect techniques to probe the framework of prion proteins. Proteinase K, a protein digester, "can be a good structural tool" for PrP^{Sc}, noted protein chemist Shu G. Chen of Case Western Reserve University.[7] The protease can dissolve PrP^C entirely, but it can only partially disassemble PrP^{Sc}. (As a result, some researchers prefer the term PrP-res, for protease-resistant prion protein, because it focuses on the observed phenomena rather than the assumed behavior as a disease agent, as the term PrP^{Sc} connotes. In this vocabulary, PrP^C is sometimes called PrP-sen, for protease-sensitive prion protein.) What's left after the proteinase K treatment is an indigestible residue of the PrP^{Sc} molecule, the size of which depends on the type of prion disease and the genetic makeup of the victim. In other words, the size of the protease-resistant residue depends on the sequence of amino acids of the prion protein. Chen and his colleagues used this approach to show that prion diseases can be grouped broadly into two types.[8]

More specific analyses of proteins require crystallization—that is, the process by which the atoms lock into place to form a solid with a regular, repeating structure. Once you have this solid crystal block, you can probe the sample in several ways. If you send x-rays through, the atoms scatter the radiation and produce a pattern that indicates the atoms' places in the crystal structure, thus revealing the molecule's construction. Scientists can also perform nuclear-magnetic resonance spectroscopy (NMR)—what amounts to an MRI scan on the protein—and create models of the molecule.

Studies using NMR and other methods have confirmed that the main structural differences between the normal prion protein PrP^C and the lethal version PrP^{Sc} is conformational: They differ only in the way they are folded. Newly made proteins from ribosomes appear as loose windings of string. But in less than a second, each protein adopts a unique roller-coaster configuration, incorporating loop-the-loops, hairpin turns, gentle bends, and rippled straightaways. A protein could not function properly without these features.

In PrP^C, three areas of the molecule twist like a right-handed screw. These so-called alpha helices, a common feature of proteins, make up around 40 percent of the PrP^C molecule. Only about 3 percent of it consists of so-called beta sheets, which are flattened-out areas that resemble corrugated tin roofs. In contrast, PrP^{Sc} consists of about 20 to 30 percent alpha helices and 40 to 50 percent beta sheets. The high beta sheet content isn't that surprising—researchers noted that the amyloids of other diseases contain many beta sheets, often stacked one on top of the other.

The question that scientists still struggle with is, How do some of the alpha helices of PrP^C turn into the beta sheets of PrP^{Sc}? The two forms of PrP definitely seem to be interacting in some way. "In vitro experiments have shown that when PrP^C is taken out of the context of membranes, it can bind selectively to PrP^{Sc} and be converted to a protease-resistant state indistinguishable from that of PrP^{Sc} itself," wrote Byron Caughey in a review that asked if prion proteins interacting with one another is the "kiss of death."[9] Experiments more closely mimicking the cellular environment found similar results, and the amount of conversion correlates with disease infectivity.

Crystallographic and NMR studies may have revealed the "lips" of the deadly kiss—that is, the area of the PrP^C molecule that first begins to change. In a test-tube solution of PrP^C, Witold K. Surewicz of Case Western Reserve University, Vivian C. Yee of the Cleveland Clinic Foundation, and their colleagues managed to spy a "dimer" of PrP^C— two individual PrP^C molecules linked together. Usually they are seen in the lab as individual molecules, although scientists have speculated that prion proteins in cells can naturally form dimers. The researchers found that an alpha helix of one PrP^C molecule likes to wrap around that of the partner PrP^C. The region where they interlock may be the place where a beta sheet first forms and may explain how PrP^{Sc} can grab hold of PrP^C.[10] "We speculate it might be important," Surewicz said of the team's work, reported in September 2001. It might be the intermediate step of the PrP^C's conversion to PrP^{Sc}. "But the fact is, we don't know," he admitted.[11]

The problem is the insoluble, aggregated, polymerized form of PrP^{Sc}. "You cannot crystallize them," Surewicz explained, meaning that the PrP^{Sc} molecules refuse to fall into discrete positions and therefore preclude the use of high-resolution studies. "Low-resolution approaches will tell you how many alpha helices and what percentage of beta sheets, but the estimates are gross," Surewicz said.

No one has been able to watch PrP^C directly interact with PrP^{Sc}, so instead researchers have used what they do know to come up with a couple of guiding theories. One, advocated by Stanley Prusiner, is the template-directed model. It postulates that PrP^C can exist in a stable, intermediate state somewhere between its normal and pathogenically folded state. Called PrP*, the intermediate form then interacts with a different protein, which Prusiner dubbed Protein X. As a result, PrP* is able to bind with PrP^{Sc}, forming a dimer. PrP* then spontaneously adopts the beta-sheet-dominated shape of PrP^{Sc}. The two split apart and go on to recruit other PrP^C molecules.

The more widely accepted working hypothesis is the nucleated polymerization model—basically, a chunk of PrP^{Sc} serves as a seed. "The seeding hypothesis," Charles Weissmann explained, "says that the infectious agent is really an assembly of molecules—simply, a crystal. So the idea is that, depending on the structure of the crystal, the molecules that add to it will adapt to whatever the conformation is."[12] In this model, single molecules, or monomers, of PrP^{Sc} lock together with other PrP^{Sc} molecules. PrP^C would bind to this aggregate and adopt the PrP^{Sc} form. The Lego-like construction keeps going, and the polymerization of PrP^{Sc} ultimately yields the scrapie-associated fibrils (prion rods) that Patricia Merz had spotted under the electron microscope. That the rods remain microscopic in length suggests that the aggregation does not go on indefinitely. At a certain point the polymer is cleaved—perhaps by "chaperone" proteins or because the structure becomes mechanically unstable. In any case, once the PrP^{Sc} polymer fractures into smaller lengths of PrP^{Sc}, they can go on to recruit more PrP^C.

In fact, this kind of crystal seeding occurs throughout the physical world. "In chemistry," Weissmann said, "there are examples where one and the same compound can crystallize in two or more forms. Once you have crystallized in a particular form in the lab, you will find that things

will not always crystallize that way, because the air is full of tiny particles which fall into the solution and seed crystallization. In another lab,
it could be a different crystal form." Such variability in crystallization
explains why no two snowflakes are ever alike; each frozen drop experi-
ences unique temperature and humidity conditions as it grows in the
cloud and falls to the earth. Other evidence to support the nucleated
polymerization model comes from yeast prions, in which Lindquist,
Caughey, and others proved that yeast prions propagate by this seeding
method. The amyloid fibers of Alzheimer's disease also grow in this
manner.

The actual conversion process seems to take place within cells, based
on experiments using cells cultured in a dish. Normal prion protein,
PrP^C, is made by ribosomes, which sit on an organelle called the endoplasmic reticulum. This structure feeds the newly made PrP^C molecule
to another organelle, called the Golgi complex, which modifies the protein and directs it to the cell membrane. The PrP^C molecule exits the
cell but remains anchored to the membrane via a molecular tether. A
cell membrane will sometimes pinch inward to bring in material from
the outside; the ends touch and fuse, forming an interior bubble called
an endosome that moves deeper into the cell. If there's PrP^{Sc} floating
around outside the cell, it can get incorporated into the endosome,
where it may react with the PrP^C lining the membrane (now the interior part of the endosome). Eventually, the endosome returns to the rim
of the cell and fuses with the cell membrane—the bubble "pops" and
spills its contents of converted PrP^{Sc} out into the extracellular space.
Alternatively, PrP^{Sc} may build up sufficiently to explode the cell. In
both cases, the freed PrP^{Sc} then goes on to recruit other PrP^C.

Although plenty of experimental evidence supports the idea that
PrP^{Sc} is the source of infection and propagation of spongiform
encephalopathies, curious and notable oddities permeate the findings.
In 1995, for instance, in vitro tests conducted by Caughey and Peter
Lansbury of Harvard Medical School showed that malformed prion
protein converts the normal one, but the amount of converted PrP^{Sc} is
less than or at best equal to the amount of PrP^{Sc} added. The conversion
wasn't self-propagating, as one might expect from the inexorable
course of TSEs and the continuous buildup of PrP^{Sc} in the brain.

More important, the researchers couldn't prove that the newly
formed PrP^{Sc} was actually infectious. If the protein-only hypothesis is
correct, then a synthetically made prion should cause disease as effi-

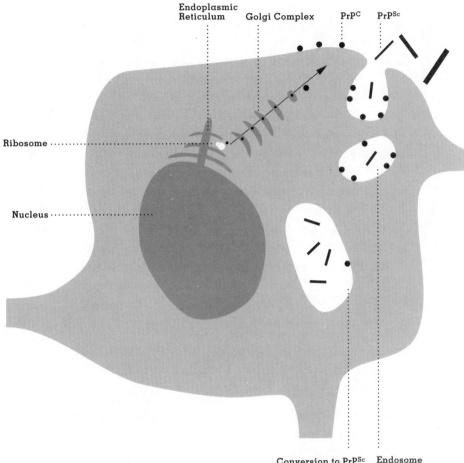

The cellular trafficking of the prion: The prion protein (PrP^C) emerges from a ribosome on the endoplasmic reticulum. The protein makes its way through the Golgi complex and onto the surface of the cell membrane. The PrP^C molecules are brought back into the cell when the cell membrane pitches in and fuses to form a bubble called an endosome. The endosome may take with it any pathogenic prions (PrP^{Sc}) floating outside the cell. PrP^{Sc} then converts PrP^C to PrP^{Sc}.

ciently as one extracted from infected animals. But experiments attempting such a proof have failed. For instance, John Collinge and his colleagues at the Prion Unit of the Institute of Neurology in London created PrP^C via recombinant means (produced through genetic engineering). Mixed with PrP^{Sc} in a test tube, some of the PrP^C became protease-resistant, suggesting a conversion to PrP^{Sc}. The researchers then inoculated the newly converted prion into mice genetically engineered to be naturally susceptible to that strain of PrP^{Sc}. They examined these transgenic mice twice weekly for signs of neurological disease, but the animals remained free of any TSE for more than 550 days (the length of the experiment). In contrast, the incubation period for the same breed of mice inoculated with the natural scrapie strain is about 180 days. Charles Weissmann informed me of a number of other transgenic experiments, in which a gene for PrP^{Sc} was introduced into cells such as lymphocytes; none of the cells were able to propagate infectivity. "The bottom line," he said, "is that PrP is essential but not sufficient."

Cofactors or Cold Fusion?

The inability to create infectivity from a test-tube preparation of PrP^{Sc} suggests two things. One is that some other compound, present in vivo (in the animal), helps create the infectivity—that is, a cofactor helps fold PrP^C into PrP^{Sc}. Indeed, the cell relies on several types of so-called chaperone molecules to direct newborn proteins to fold properly. The yeast prion [PSI] relies on a chaperone molecule called heat-shock protein 104 to fold it from one form to another. (Besides helping proteins to fold, heat-shock proteins are also produced by cells as a means of coping when they are stressed by, say, heat.) Prusiner has referred generically to the essential cofactor for producing PrP^{Sc} as Protein X, but no one has had any success in identifying it. PrP in both its healthy and deranged forms can bind to many types of molecules, including glucose polymers, metal ions, lipids, and nucleic acids, thereby making the identification of a chaperone molecule difficult.

A good candidate for a conversion chaperone is sulfated glycosaminoglycans (GAG), one of the most abundant class of sugar molecules in the body. These polysaccharides line many cell membranes and occupy the spaces between cells; they structurally support cells and lubricate joints (the most common is chondroitin sulfate, the stuff of cartilage). For many years, researchers have known that GAGs were associated with protease-resistant PrP and tend to coat PrP^{Sc} in the brain. Researchers in Caughey's lab wanted to see what the direct effect of these molecules might be in the conversion reaction. What they found was surprising. The GAG molecule heparin sulfate "can serve as quite a profound stimulator of the conversion reaction"—boosting the change from PrP^C to PrP^{Sc} by a factor of five to six, Caughey explained.[13] The finding was puzzling, because researchers had found that in tissue-culture and in vitro experiments, added GAGs could *inhibit* the formation of PrP^{Sc}.

To explain the contradiction, Caughey speculated that GAGs or GAG-like molecules do like to bind to PrP^C. But only when the GAGs are from the host do they boost the conversion to PrP^{Sc} (if the pathogenic form is present). Adding GAGs from an outside source may be inhibiting PrP^{Sc} formation by binding to the PrP^C before the body's own GAGs can do so.

The second conclusion that one can draw from the inability of test-tube-made PrP^{Sc} to infect healthy animals is that the prion theory is wrong. There are several mavericks who think that nucleic acids, not proteins, might cause TSEs. One of the most thoughtful on the issue is Robert G. Rohwer, director of the molecular neurovirology laboratory at the Veterans' Affairs Maryland Health Care System in Baltimore. He is so careful, one researcher told me, that he often doesn't publish because he keeps coming up with counterarguments to his own conclusions—and counterarguments to those counterarguments. He is also famous in the field for having once referred to the prion hypothesis as the "cold fusion" of infectious disease.[14]

Rohwer, who began his TSE career when he joined Gajdusek's group in the mid-1970s and published his first CJD paper with him in 1977, claims that the main arguments used to swing thinking toward the protein-only camp do not actually do away with a virus. In a paper he wrote in 1991 entitled "The Scrapie Agent: 'A Virus by Any Other Name',"[15] he systematically cataloged his objections, based on his own and others' work. He reexamined the data from Tikvah Alper, who had found that the scrapie agent survived blasts of radiation and had concluded that it was too small to be a virus. But the "target theory" calculation that Alper used to determine the scrapie agent's size makes no distinction among viruses, enzymes, and other molecules, Rohwer stated. A virus' sensitivity to radiation depends on whether it has a single or double strand of nucleic acid—single-stranded viruses, such as the yellow fever and tobacco mosaic viruses, survive radiation blasts better because, being smaller, they are less likely to be hit. Using empirical data on such viruses, Rohwer calculated that the scrapie agent's radiation resistance isn't all that different from very small viruses, such as the porcine circovirus, which targets the immune system of pigs.

He also dissected the inactivation studies—how temperature and bleach solutions, for instance, affected the infectivity of the scrapie agent. His experiments showed that the kinetics of the inactivation— how infectivity changes over time and under various conditions—were similar to those of viruses. In fact, 99.99999 percent of the scrapie agent population is killed in the minute when the temperature shoots from $4°$ C to $121°$ C. Granted, some infectivity remains—but that can also happen for viruses. "There are subpopulations resisting complete disinfection," Rohwer stated.[16] In fact, "that was part of the original polio vaccine story." Some vaccine recipients injected with the "inactivated" polio virus actually came down with the disease. "You could never get rid of that last little bit. And there was no way to prove that you got rid of it without using up the entire vaccine stock," Rohwer explained. It's a problem that still dogs those trying to guarantee that vaccines will not trigger the diseases they're trying to prevent.

Rohwer conducted many of his experiments in the 1980s, when he was seeking to determine whether "this stuff is outside the normal paradigms of molecular biology," as he put it. Based on the experiments he has been able to do—sacrificing thousands of animals to generate good data is quite expensive—he has concluded that "these things don't fall out of the range of classical virology." In other words, the TSE agent may not be a prion protein at all but an unusually small, ubiquitous virus—or maybe even the hypothetical virino.

But just as prion advocates haven't definitively proved the protein-only theory, so too have the nucleic-acid proponents failed to show a single A, T, C, or G that might be part of the cause of TSEs. Certainly the technology exists to find the agent, even if it is made of only a few nucleotides. "What's lacking is the will to do it," believes Rohwer, who funded his experiments out of his own pocket, thanks to profits from commercial studies done in his lab. Indeed, prion doubters complain that it's difficult to get funding to study TSEs from government sources without mentioning "prion" in the grant applications. Rohwer recalled putting in a proposal to the NIH in the mid-1980s to search for nucleic acids of the scrapie agent. The rejection letter came back with comments that effectively told Rohwer to, in his words, "get with the program, you know, we're doing prions now."

Then, in 1997, the Nobel committee awarded Stanley Prusiner the prize for a "new biological principle of infection." That didn't please skeptics, including Yale University neuropathologist Laura Manuelidis and virologist Bruce Chesebro of the NIH Rocky Mountain Labs. They expressed concern that the prion/virus debate would be shunted aside completely. "It would be tragic if the recent Nobel Prize award were to lead to complacency," Chesebro wrote in the January 2, 1998, issue of *Science*. "It is not mere detail, but rather the central core of the problem, that remains to be solved."[17]

Actually, rather than marginalizing the nucleic-acid camp, Prusiner's Nobel may have had the opposite effect. The prize "focused attention on the detractors [of the prion theory] as well as the promoters, so it got the issues out more in the open than it had been," Rohwer said. He added: "I think that a lot of people who hadn't thought about it very critically suddenly had to, and I hear a lot more sympathy for an open mind on this question."

But the continued accumulation of positive data—and the lack of definite results that say otherwise—has made the prion theory ascendant. To most minds, malformed proteins are the most likely cause of spongiform encephalopathies. To the minority of TSE workers who still resist the idea, the proof of the prion hypothesis is very simple. Make the molecule synthetically, and see if it triggers a spongiform encephalopathy. "No one's been able to turn that stuff into something that's infectious," Rohwer remarked, referring to Collinge's and others' work. If someone can "create infectivity de novo in the test tube, I'll drop my objections and get with the program."

The Copper Connection

Considering that PrP^C is evolutionarily conserved—the protein has been found in all animals examined (including chickens and turtles)—it is reasonable to assume that it plays a crucial role in the body. Maybe the disease occurs because of the loss of this function. Yet curiously, despite years of work, no one knows for sure what good PrP^C is. In 1992, Prusiner, Weissmann, and their colleagues "knocked out" the prion protein gene in a line of mice. These PrP-null mice developed and acted just like their normal cousins, showing no ill effects from not having PrP^C, at least as far as you can tell from observing rodents. Whatever role PrP^C plays, the study suggested, cells have alternative pathways and can compensate for the loss of PrP^C.

PrP^C's job almost certainly involves copper. In humans, copper is an essential trace element, critical in many oxidation-reduction pathways and chemical reactions. The mineral helps the body make collagen, melanin, and hemoglobin. (*Star Trek* fans may recall that the green-blooded Mr. Spock relied on copper, rather than iron, to ferry oxygen.) It is critical in the nervous system, being essential for the production of noradrenalin, a neurotransmitter, and for the myelin sheaths around nerves. Too little, and anemia, skin sores, poor immune functioning, and even atherosclerosis may result. As is the case for other trace elements (and a few vitamins), too much copper is toxic—headaches, nausea, and potentially fatal kidney damage is possible.

PrP^C seems to have a particular affinity for copper. In 1995, Martin P. Hanshaw and his colleagues at the MRC Neurochemical Pathology Unit of Newcastle General Hospital in the U.K. conducted a test-tube study using a synthetically generated stretch of PrP^C. They found that the octapeptide region of the protein prefers to bind to copper rather than to other metals.

But the real excitement among copper-binding specialists occurred two years later in New Orleans, at an annual meeting of the Society for Neuroscience. David R. Brown of the University of Cambridge gave a presentation that sent e-mails zipping among prion researchers. Brown, who collaborated with Hans Kretzschmar of the Georg-August University of Göttingen in Germany and others, cultured neurons from normal and PrP-null mice and reported that the cells lacking PrP^C suc-

cumbed more readily to poisoning by copper sulfate than did cells with
the prion protein. Adding a segment of PrP^C that likes to stick to
copper protected the genetically altered neurons.

Moreover, in another experiment, Brown found that normal mice
had about 20 times more copper than the PrP-null mice did—PrP^C was
evidently grabbing onto copper. (Subsequent work, however, showed
much less of a divergence, about a 50 percent increase.) To Brown, the
data suggested that PrP^C could be sopping up potentially hazardous
copper ions released when neurons fire. It could also mean that PrP^C
was taking the copper to give it ultimately to enzymes that needed it.
One such enzyme is superoxide dismutase, a powerful antioxidant that
protects cells from the damaging free radicals that result from metabolism. The PrP-null cells, in fact, were less able to resist such oxidative
harm than normal cells.[18,19]

Additional evidence that PrP^C is intimately connected with copper
comes from studies of cultured neuroblastoma cells by David A. Harris
of the Washington University School of Medicine in St. Louis. In work
done in 1998, he and his collaborators watched as, in the presence of
copper, cells quickly internalized the PrP^C molecules that sat on the
cell surface. PrP^C's job may therefore be as a trafficking agent, taking
copper into and out of cells.

Still, despite the clear attraction that PrP^C has for the mineral, the
protein probably plays a minor role with respect to copper usage by
cells, Harris has concluded. Evidence for that comes from his studies
involving PrP-null mice, normal (wild-type) mice, and mice genetically
engineered to produce ten times the normal amount of PrP^C. The three
mice strains did not show any significant differences in superoxide dismutase activity or uptake of copper needed to make the enzyme, in
contrast with Brown's work. Moreover, copper levels were similar in
their brains, suggesting that PrP^C isn't the main means by which cell
membranes latch onto copper. So PrP^C's role in copper metabolism
could be rather small.[20]

Perhaps PrP^C has multiple minor roles. One was found by researchers
led by Odile Kellermann from the Pasteur Institute in Paris who
reported in 2000 that PrP^C in mouse neurons acts as signal relays. It
interacts with as-yet unidentified molecules in the synapse between neurons and then somehow triggers other chemical agents inside the cell to

act. The team also found at least two other cell-surface proteins involved in this particular signaling cascade, suggesting that PrP^C is a bit player, perhaps only designed to fine-tune the signals between neurons.[21]

Double Trouble

Although PrP-null mouse studies did not reveal the function of PrP^C as researchers had hoped, they did, quite by accident, reveal that the prion protein has a little nephew. In 1996, researchers from the Nagasaki University School of Medicine developed a strain of PrP-null mice that got sick spontaneously, in contrast to previous lines of PrP-null mice. Between 6 and 12 months of age, the Nagasaki mice developed ataxia from a loss of Purkinje cells. These specialized neurons relay signals out of the cerebellum, the brain structure that controls muscle coordination. Because the problem could be prevented by the introduction of the PrP gene, the researchers concluded that the ataxia arose because of the absence of PrP^C.

The real reason that the mice became ataxic, however, turned out to be the way the Nagasaki researchers had cut out the PrP gene of the mice. Other PrP-null mice, made by labs in Edinburgh and Zurich, were missing only about two-thirds of the protein-encoding part of their PrP gene (such a partial removal is usually enough to curtail the expression of the normal protein). The Nagasaki investigators, however, removed the entire protein-encoding sequence of the gene, as well as some of the flanking regions. Sure enough, when bigger genetic chunks of the Zurich and Edinburgh strains were similarly knocked out, they too developed ataxia.

The explanation for why the fraction of PrP removed makes a difference came from a 1999 report by California researchers that included David A. Westaway, Richard C. Moore, Leroy Hood, and Stanley Prusiner. Evidently, the Nagasaki team didn't slice out *all* of the relevant parts of the PrP gene: They had left in the switch that turns the gene on. (Technically speaking, they had left the PrP gene's promoter region intact.) Without a PrP gene to trigger, the switch activated the next gene it encountered. It turned out that this gene is extremely similar to the PrP gene. Westaway and his colleagues called it Prnd, and the protein it makes *doppel* (German for double).

Consisting of 179 amino acids, doppel is a few dozen amino acids shorter than PrP^C. They share about 25 percent of the same sequence of amino acids. Both are evolutionarily conserved, appearing in humans, mice, sheep, cattle, and probably most other animals. Both anchor themselves to the cell surface and contain three alpha helices. But the doppel gene is expressed mostly in the testes and hardly in the brain at all. Too much doppel was produced in PrP-null mice and caused disease, probably by acting like a truncated form of PrP^C. This short form, lacking the octapeptide repeats and the amino acids from 106 to 126, had been shown to cause ataxia in transgenic mice.

Much remains unknown about doppel. In healthy animals, it probably has little to do with nerve cells. When researchers knocked out the doppel gene in mice, sterility in males was the dominant symptom.[22] The connections between doppel, PrP^C, and prion disease remain an open question.

The search for the scrapie agent transformed into an investigation of the prion protein after Stanley Prusiner had published his landmark paper in 1982. In the two decades since, research into transmissible spongiform diseases exploded, and the understanding gained about the prion has convinced most scientists that a protein is the basis for a class of transmissible and heritable neurodegenerative diseases. But much of the interest—and funding for laboratory work—resulted not just from the novel idea that a protein could act like a virus. The burning cattle carcasses around the English countryside certainly had much to do with the acceleration in TSE research.

CHAPTER 8

Consuming Fears

Modern agriculture enables prions to adapt to a new host, creating the dread mad cow disease.

Healthy Friesian dairy cows do not lose weight, drool, arch their backs, and menace the other cows. Clearly, something was wrong with Cow 133, but the symptoms baffled Peter Stent, who ran the Pitsham Farm in Sussex, in southern England. He called for help, and on December 22, 1984, a local veterinarian, David Bee, arrived for what would turn out to be the first of many examinations of the eight-year-old bovine. Bee's initial hunch was a kidney infection, but an injection of 100 cc's worth of duphacycline did not improve her condition. By February 4, 1985, "she had developed head tremor and in-coordination," Bee recalled.[1] She died a week later.

In subsequent visits to Pitsham Farm, Bee and his partner, Michael L. Teale, encountered additional cows with similar symptoms. Both vets noted how aggressive the animals had become—one even chased Teale after she'd fallen to her knees. By April 29, 1985, six more cows had died. Samples from various organs were sent to the Veterinary Investigation Centre in Itchen Abbas, Winchester, one of 22 centers set up around England and Wales to conduct surveillance of new animal diseases. The tests provided no firm diagnosis, however. "I thought the likely cause was some form of toxin particular to Pitsham Farm," Bee speculated. "There was a small special brickworks in the centre of the farm, and I supposed that the source of toxin may lie there. However, tests for lead and further tests for mercury were all negative."[2] The vets looked for evidence of parasitic infection and poisoning. They even looked at a dead magpie and her eggs to check for toxic compounds. But they found nothing conclusive. The closest they got to some kind of diagnosis came after they noticed fungal growth in the feed hopper supplying the milking parlor.

At the vets' urging, Stent agreed to sacrifice one of his sick cows, and on September 2, 1985, he sent Cow 142 to the VI Centre. She was euthanized, her brain removed, soaked in formaldehyde, and sent for analysis to the Central Veterinary Laboratory (CVL) in Weybridge, Surrey, which acts as a focal point for all the U.K.'s veterinary activity. (It is now called the Veterinary Laboratories Agency.) The on-duty pathologist, Carol Richardson, examined the wrinkled organ 11 days later. In two areas of the brain and the spinal cord, she recorded mild to moderate "neuronal and neuropil vacuolation"—some sort of spongiform encephalopathy had taken its toll.[3] But if there was a suggestion that this was scrapie in a cow, it was never recorded. Instead, the pathologist's write-up suggested that, most likely, a toxin had destroyed the tissue. Once the CVL report was completed in October 1985, the matter was dropped. "By this time," Bee stated, "new cases had ceased to developed. I imagined that the problem had run its course."[4]

In fact, while Bee was looking at the uncoordinated, hostile cows of Pitsham Farm, another vet, Colin Whitaker, was seeing similar symptoms in other bovines.[5] On April 25, 1985, he was called to examine a Friesian-Holstein cow on Plurenden Manor Farms in Kent, in southeastern England. What was once a tame, quiet animal had become a staggering, aggressive nuisance. But Whitaker could find nothing wrong except for some ovarian cysts, which he treated. Her behavior continued to change for the worse, and Whitaker suggested that she might have a brain tumor. Three months later, on July 26, 1985, the cow was put down and buried.

Whitaker would visit the manor again and again over the next year, as the stockmen continued to find cows with the same symptoms. By the end of 1986, seven cows were destroyed because of this mysterious disease. In November 1986, Whitaker consulted with the Veterinary Investigation Centre in Wye. The vet there shipped the brains of three sick bovines to the CVL, where Gerald A. Wells examined the tissue under the microscope. "Each of the three Wye cases I examined had a common novel pathology, with the essential change being a 'multifocal spongy transformation' of the brain and a degeneration of neurons in the brain stem," Wells recalled. "Compared with most other animal disorders the changes most closely resembled scrapie, but there were subtle differences." That, coupled with a similar brain sample from a different herd, "indicated that this might represent the onset of an epidemic," Wells surmised.[6]

Because other conditions, such as inflammation or poisoning, can occasionally fill the brain with microscopic holes, the CVL was careful about circulating the information. On December 19, 1986, Raymond Bradley, the department head and Wells's boss, wrote a confidential memo to William Watson, the CVL director, and Brian Shreeve, the CVL director of research, alerting them to the pathological findings:

> I would advise keeping an open mind about the aetiology until we have more information. The principal lesions are degenerative and non-specific. If the disease turned out to be bovine scrapie it would have severe repercussions to the export trade and possibly also for humans if for example it was discovered that humans with spongiform encephalopathies had close association with the cattle. It is for these reasons I have classified this document confidential.[7]

News of the potential bovine scrapie soon reached the Ministry of Agriculture, Fisheries, and Food (MAFF), a government department analogous to the U.S. Department of Agriculture. There, a staff official saw William Rees, the chief veterinary officer, "walking down the passage with steam coming out of his ears" after hearing the news.[8] Rees understood the implications of bovine scrapie.

Word of more cases from different herds was arriving at the CVL, and by February 1987, CVL virologists had spotted the scrapie-associated fibrils (SAF) in homogenized brain samples. Bradley asked Wells to prepare a draft for a possible article on the clinical signs and pathology for *Vision*, a newsletter circulated among the Veterinary Investigation Centres. Wells resisted, arguing that publicity at this point was premature and that they should wait for further data. Nevertheless, he produced a draft by early March, but because of his and another vet's objections, Bradley and CVL director Watson decided not to push for its immediate publication.

At Watson's request, however, Wells did present the information on the afternoon of May 29, 1987, to a closed-door meeting of the Medical and Veterinary Research Clubs held at St. Bartholomew's Hospital in

London. The meeting involved members of the scrapie and human TSE communities, including Richard Kimberlin of the Neuropathogenesis Unit in Edinburgh. "Although there was significant scientific interest," Wells stated, "as far as I was aware, no one at the meeting expressed the view that the disease would develop into a 'calamitous' epidemic in the U.K."[9] Actual public notification would not occur until October 31, 1987, when a two-page communication was published in an issue of the *Veterinary Record*: "A Novel Progressive Spongiform Encephalopathy in Cattle"—a disease that Wells, Bradley, and their co-authors dubbed bovine spongiform encephalopathy, or BSE.[10]

Tracking the Source

With more farms reporting cases of "mad" cows, MAFF decided in June 1987 that it was time to figure out the cause and enlisted John W. Wilesmith of the CVL. Wilesmith had become head of the epidemiology department a year earlier, after joining the CVL in 1976. His only contact with TSEs had come from a few cases of scrapie that he had seen in the early 1970s, during his time as a vet student and private practitioner.

Following a meeting with Wells on June 3, 1987, Wilesmith recounted, "I began to formulate my approach and objectives for an initial epidemiological investigation. I decided that I should consider all possible causes of the disease and obtained the identities of herds in which cases had been confirmed" or suspected.[11] He sent out questionnaires to these herdsmen, retrieving various details about the cow, such as sex, breed, date of birth, pedigree, origin, identities of the offspring, and date when symptoms appeared. Wilesmith and his co-workers also gathered data on the herd, such as its size and proximity to sheep (several farms kept both types of livestock) and feeding practices, as well as the types of pharmaceuticals and pesticides used. The team also constructed a computer model to determine the onset of the burgeoning epidemic. "Although infection was undoubtedly the cause of BSE," wrote Richard Kimberlin in a review, "it was important to eliminate other possibilities, particularly as the transmissibility of BSE had not been demonstrated at the time."[12] Transmissibility would be proved in

October 1988, when Edinburgh researchers at the Neuropathogenesis
Unit reported that mice could contract BSE via intracerebral inoculation.

Wilesmith, Wells, and their colleagues discovered that BSE occurred
predominantly in dairy herds: From April 1, 1985, to March 31, 1988, the
incidence in dairy cattle (311 out of 44,767) far outnumbered those in
beef herds (11 out of 54,166).[13] The primary breed affected, Friesian-
Holstein cattle, was no surprise, given that this breed made up about 90
percent of the dairy cattle in the U.K. All victims were adults, ranging in
age from 2.75 to 11 years, and the incidence within herds ran from 0.2
percent to 11.1 percent, with an average of 1.52 percent. The onset of
symptoms did not vary with the calendar month or pregnancy stage, as
might happen if the illness were associated with the seasonal use of vaccines, herbicides, or pesticides. The fact that the disease appeared
simultaneously in widely separated areas of the U.K. argued against a
single source of infection, such as the sudden emergence of a mutant
scrapie strain that could infect cattle, which would have produced cases
radiating from one area. And BSE didn't seem to spread the way scrapie
in sheep did—there was no evidence of transmission from individual to
individual or from mother to calf. "The features of note were that all
cases appeared to be index cases and the form of the epidemic was typical of that within an extended common source," recollected Wilesmith.[14]

The only factor common to all cases, the team found, was the dietary
supplement regularly fed to dairy cattle beginning at weaning. The pro-
tein-rich feed ensured rapid growth and high milk production. It was
made from the meat and bones of dead animals—and very likely contained material from sheep that succumbed to scrapie. The team
hypothesized that something in the production of meat-and-bone meal
(MBM) allowed a scrapie-like agent to contaminate the cattle feed. If
true, the disease had been triggered by a world event that had no immediately obvious food-safety implications: the decision of the
Organization of Petroleum Exporting Countries (OPEC)—founded in
1960 by Iran, Iraq, Kuwait, Saudi Arabia, and Venezuela—to raise crude
oil prices.

Forced Cannibalism

In October 1973, in retaliation for Western nations' support of Israel against its Arab neighbors in the so-called Yom Kippur War, OPEC decided to raise oil prices 70 percent. OPEC jacked the prices up again in December, this time by 130 percent, and imposed a temporary embargo on shipments to the U.S. The long lines at the gas pumps that many Americans endured were just part of the economic stagnation that much of the world began to experience because of the price shocks. OPEC, which now counts a dozen member countries, announced several other hikes; by 1980, the price of a barrel of crude reached $30, up from $3 in 1973.

To save on fuel costs, oil-importing nations implemented conservation measures throughout the 1970s. In the U.K., one such measure involved the production of animal feed.

Feeding cattle, sheep, dogs, and cats the cooked and ground-up remains of other cattle, sheep, dogs, and cats is horrifying to many people. After all, it's forced cannibalism; worse, cows and sheep are vegetarian by nature. Rendering—as the practice of converting carcasses into something useful is known—was upsetting enough to Oprah Winfrey that, on the air in April 1996, she swore off hamburgers. (The show briefly dragged down the price of cattle futures on the Chicago Mercantile Exchange and also brought two highly publicized lawsuits against Winfrey and one of her guests, former cattle rancher Howard Lyman, by a Texas feedlot organization. The organization claimed that the two of them had made false, slanderous, and defamatory statements. Winfrey and Lyman won the first lawsuit, and a federal judge dismissed the second in September 2002.)

However shocking it may seem today, the practice of rendering has existed for centuries. Butcher and boil the carcass, skim off the creamy white fat that floats to the top (called tallow), and you can make candle wax; add ash to the tallow, heat, and you have soap. The heavier protein sinks to the bottom, producing "greaves" that can be fed to animals. In modern agriculture, the primary source of the animal material used in feed is slaughterhouse refuse, or "offal." Offal is made from animal parts that people generally don't eat, such as the bladder, diaphragm, udder, head, hooves, and bones. (Estimates vary as to how much of a cow ends up being offal—anywhere from about one-third to almost one-half.) Renderers also take in carcasses deemed unfit for human consumption, "downer" animals on farms (those that can't walk on their own for whatever reason), as well as dead pets and roadkill. In 1988, U.K. rendering plants took in 1.3 million metric tons of animal matter: 44.8 percent came from cows, 15.3 percent from sheep, 20.9 percent from pigs, and the rest from animals of other species.[15]

In the U.K., World War II stimulated improvements in the rendering process so that more of the carcass could be recycled. One method, which became popular after the war, used solvents to get more of the valuable fat out of the offal.

"The theory of solvent extraction was to remove tallow from meat and bone greaves which could contain anything up to 30 percent tallow," explained Edward Wyatt "Bill" Bacon, a council member of the U.K. Renderers' Association who started working at his family's Birmingham rendering business in 1946.[16] Tallow fetched a higher price than greaves, making the added expense of solvent a good investment. During this process, the greaves were first crushed and then heated to just below the boiling point of the solvent, maybe around 65° to 70° C. A solvent, such as benzene, petroleum spirit, hexane, or perchloroethylene (dry-cleaning fluid), was pumped in two or three times while the solution was heated. This process took about eight hours. Next, the resulting mixture of tallow and solvent was sent to another cooker, which generated temperatures of 105° to 120° C for 45 to 60 minutes. At this stage, the mixture reached 90° C—hot enough to vaporize the solvent, which would later be condensed and reused. The tallow was recovered, and the greaves left behind were steam-blasted for 15 to 30 minutes to remove any residual solvent. The entire process could remove all but 1 percent of the tallow in the greaves.

By the late 1970s, however, solvent extraction had fallen out of favor in the U.K. because the price of tallow had dropped in relation to that of greaves. At this time, animal feed manufacturers started leaving more tallow in the greaves. The richer meat-and-bone meal (about 10 to 12 percent fat) provided additional calories to cattle. Besides, working with the volatile solvents at high temperatures posed a serious worker-safety issue. In fact, most of the solvent extraction plants in the U.S. had "blown up, burned down, or closed for safety" by 1970, according to a 1998 memo from the U.K.'s National Renderers' Association.[17] (By Bacon's estimate, U.S. rendering technology was 10 to 15 years ahead of the U.K.'s.)

And, with the rising price of oil triggered by OPEC, it just did not make sense, economically speaking, to keep using the increasingly expensive solvent and to continue with an energy-hungry process that required hours of heating. Instead, spinning the greaves in a centrifuge and pressing the material could remove a sufficient amount of fat. The percentage of meat-and-bone meal made from solvent extraction dropped from about 65 percent in 1977 to around 10 percent by 1982. Two large plants, accounting for 26 percent of meat-and-bone meal production, ceased their solvent extraction between 1980 and 1981—about the time Wilesmith and his colleagues figured that the first exposure to the infectious BSE agent occurred.

Wilesmith hypothesized that the move away from the solvent-extraction process eliminated two important treatment steps that could have reduced the amount of the scrapie agent (the misfolded prion protein in Stanley Prusiner's theory) that may have been in the mix because of infected sheep. "These [steps] were the direct action of the solvent and the additional heat treatment, including the application of superheated steam, necessary to remove the solvent. In addition the solvent had an indirect effect by reducing the lipid content, which would enhance the effect of the heat treatment," he concluded.[18] Coupled with the boom in the sheep population from 22 million in 1980 to about 35 million by 1988 (the U.K. has the highest proportion of sheep to cattle in the world) and a scrapie rate of 2.25 cases per 1000 sheep, it's very likely that a great deal of scrapie-infected sheep carcasses made it to the renderers during the 1980s.

(Initially, Wilesmith had also concluded that the change from cooking offal in batches to processing it continuously, by adding new material as the old was cooked, also played a role. He later changed his mind when subsequent analysis showed that the temperature differences between batch and continuous processing were not great and that the material in continuous systems tends to be ground smaller, which would have allowed the heat to more effectively inactivate any lurking PrP^{Sc}.)

Not all scientists agreed with Wilesmith's hypothesis. In 1990, microbiologists Richard Lacey of the University of Leeds and Stephen Dealler, a consultant with the Burnley General Hospital, pointed out that the epidemic could have started from a mutant scrapie strain that jumped to cows or even from a spontaneous mutation in the cattle's PrP gene, much the way that one in a million humans contract Creutzfeldt-Jakob disease. They considered the evidence for BSE arising from existing scrapie strains as nonexistent.

In fact, there is some evidence arguing against the idea that changes in rendering methods enabled scrapie to emerge as BSE. In an experiment carried out in Texas, cows inoculated with scrapie developed lesions that did not resemble those found in BSE-infected cattle. Experiments with rodents also suggested that scrapie was completely different from BSE: Scrapie-susceptible hamsters inoculated with extracts from BSE cows remained healthy, whereas mice that tended to resist scrapie came down with a prion disease after inoculation with BSE material.

Experiments carried out by David Taylor of the Neuropathogenesis Unit in Edinburgh further complicated matters. In 1990, in collaboration with the rendering industry, he constructed a pilot-scale facsimile of rendering plants to compare the different processing methods and their ability to inactivate the scrapie agent. Taylor used heptane to mimic the solvent used in old rendering plants. In December 1992, Taylor began injecting the meat-and-bone meal made in the facsimile plant into mice to test for infectivity. The experiments showed that the solvent-extraction process couldn't eliminate the scrapie agent: the mice still got sick. At best, the heat involved in the processes reduced the infectivity by a factor of 10.

Yet the elimination of this 10-fold reduction in infectivity might have been enough to tip the scales. Gabriel Horn, emeritus professor of zoology at the University of Cambridge, led a distinguished committee of U.K. experts investigating the origins of BSE. In a July 5, 2001, report, the group concluded that:

> Rather than switching from a situation where no TSE infectivity passed through the rendering system to one where some infectivity passed through and an epidemic ensued, it could be that a threshold level of infectivity was breached. Below this threshold a certain amount of infectivity survived the rendering process, but not enough to sustain an epidemic; above the threshold (the situation that was perhaps reached after the changes in rendering) enough infectivity survived the rendering process to initiate and then sustain an epi-

demic. Such threshold behaviour is typical in epidemics of infectious disease.[19]

Horn's panel cited a statistical analysis done in 2000 that showed that each BSE cow, after being rendered into meat-and-bone meal, infected ten other cows on average. So a ten-fold increase in infectivity resulting from the shift away from solvent extraction presumably helped to drive the epidemic once it started.

Moreover, Horn's panel found that Britain was alone among European countries in feeding meat-and-bone meal to dairy calves as part of their starter rations, switching over from reconstituted milk, feed, hay, vegetables, and fish. Australian farmers also incorporated MBM into dairy calf feeds, but notably, Australia is scrapie-free, having managed to eradicate the disease after its first outbreak in 1952. (U.S. farmers relied on cheap and abundant soybeans for the calves' protein.)

Most scientists now subscribe to the theory that BSE originated from a strain of scrapie endemic to the U.K., although probably no one will ever know for sure. All researchers agree, however, that the recycling of cows to feed other cows amplified the BSE agent. Seven out of ten calves fed just 1 gram of BSE-infected brain tissue — an amount the size of a couple of peppercorns — died from BSE. A growing calf consumes about 2 kilograms of feed every day, containing about 90 grams of meat-and-bone meal.

Determining the origin of BSE is more than an exercise in theoretical probabilities. It helps pinpoint both the start of the epidemic and the number of infected but asymptomatic cows that might have entered the human food chain. If cows were infected at calfhood, they would appear to be healthy at slaughter, which generally occurs at two years of age — far less than BSE's incubation period, which Wilesmith had concluded was about five years. If the initial cases in 1985 resulted from meat-and-bone meal derived from the carcasses of BSE cows instead of scrapie sheep, then the first infections must have occurred before 1980 and went undetected. If so, then tens of thousands of infected cows became human food before 1985.[20] And each cow could have infected 500,000 people.

More than two centuries of living with scrapie convinced some Britons that the possibility of humans contracting BSE was unlikely. After all, review after epidemiological review showed no connection between human prion diseases and scrapie. But given that the infectivity of the prion protein can shift when passing from one species to another, there were no guarantees.

By the end of 1987, there were 446 BSE cases in the U.K.

Tackling an Epidemic

At least some officials at the Ministry of Agriculture, Fisheries, and Food were worried enough about BSE to propose safeguards that would protect both human and bovine populations. On February 24, 1988, MAFF's Permanent Secretary, Derek Andrews, suggested to the Minister of Agriculture, John MacGregor, that Britain pursue a policy of compulsory slaughter and compensation with farmers. Unsure, the minister turned to Sir Donald Acheson, the Department of Health's Chief Medical Officer. Acheson recommended that an expert panel be set up to advise on BSE's human health implications. After consulting with other department ministers, Acheson wrote to Andrews that all were behind the formation of the expert panel. They favored, Acheson wrote, "a low-key announcement by MAFF, in view of the fact that it is important that the public should not be given a false impression that a health risk in man is likely."[21]

The result was a working party headed by Sir Richard Southwood, a zoologist at Oxford University and chairman of the National Radiological Protection Board. The Southwood Working Party met for the first time on June 21, 1988.

Just a week before, the U.K. took its first measure to stem the burgeoning epidemic. A ban on feeding ruminant-derived feed to other ruminants was declared, effective July 18, 1988. That meant no feeding of cattle to cattle, or sheep to cattle, or cattle to sheep, or sheep to sheep. Given that Wilesmith suspected that meat-and-bone meal was the probable source of the infection, many have wondered why it took seven months to impose a ban. The delay may in fact have been initiated by the U.K. Agricultural Supply Trade Association, which asked for a grace period of three months; that way, their members could clear existing stocks of feed. After all, if cows had been consuming massive doses of infected feed for at least the past six years, what was another few weeks? The trade group and even some MAFF officials remained skeptical of the hypothesis that meat-and-bone meal was the vector and thus felt no urgency in imposing a ban, which was initially intended to be temporary and was to be lifted at the end of the year. After its first meeting, the Southwood committee made its first interim recommendation: Slaughter all cattle suspected of being infected with BSE. Carcasses were thrown into quarries, doused with gasoline, and torched. The gruesome sight of the bovine pyres, which drew objections from nearby residents, and the increasing numbers of cows that had to be destroyed soon led to enclosed incineration. Farmers were compensated at 50 percent of the animals' market value if a postmortem revealed BSE, at 100 percent if not. (The compensation for BSE cattle was raised to 100 percent in February 1990.) By November 1988, the committee recommended that the ruminant feed ban be extended indefinitely. (In 1994, U.K. authorities banned all mammalian protein from cattle feed and, two years later, from all animal feed.)

The Southwood Working Party met three more times and delivered its final report to the U.K. ministers on February 9, 1989. The committee endorsed Wilesmith's conjecture that scrapie-infected sheep offal was part of the meat-and-bone meal fed to cattle in the early 1980s and that it had caused BSE. Based on their knowledge of scrapie, the Southwood committee concluded that it was "most unlikely that BSE will have any implications for human health" and that "the risk of transmission of BSE to humans appears remote."[22]

The committee did not think that additional steps were necessary to deal with subclinical BSE cases—infected cows that appeared healthy. The committee, however, advised baby food manufacturers not to include offal such as the thymus in their products, although the liver and kidneys were fine. The committee gave infants extra protection for several reasons. First, baby food was homogenized and thus is more likely to contain high-risk organ meat. (In fact, British fashion in the 1930s dictated that infants should be fed sheep brains.) Babies also have underdeveloped immune systems and are naturally prone to more infections, at least from conventional pathogens. Finally, there was some evidence that calves and lambs were more susceptible to the BSE and scrapie agents than adult animals. Later research showed, however, that babies are no more likely than adults to contract TSEs.

As a precaution, millions of cattle were slaughtered and incinerated—here, at Cluttons Animal By-Products in Wrexham, Wales, U.K. (*Nigel Dickinson/Peter Arnold, Inc./Still Pictures.*)

Years later, when he testified as part of a public inquiry investigating the handling of the BSE epidemic, Sir Richard recounted that:

> We were very particular about the wording of paragraphs; and that we did not want it to be too reassuring. We wanted to point out that there were enormous uncertainties. And that if these uncertainties turned out to be more likely than we had judged there could be catastrophic and very profound consequences.[23]

Knowledge about TSE was limited at that time — Stanley Prusiner was still catching flak for his prion hypothesis — and the Southwood Working Party members admitted that because of the limited data, their risk assessment could be wrong. And if so, the consequences could be grave.

Any sense that the risk to humans *could* be high, however, did not make an impression on the British authorities. The BSE public inquiry later concluded:

> Unhappily, the *Southwood Report* was treated by many officials in the Ministry of Agriculture, Fisheries and Food (MAFF) and the Department of Health (DH) and, at times, by Ministers as if it contained definitive conclusions based on an evaluation of adequate data.[24]

Moreover, the focus on baby food may have given false assurances that adults could not get sick from BSE.

While the British government optimistically concluded that BSE posed no threat to humans, the media presented a radically different view. The day the Southwood report was released, a neuropathologist on a BBC television broadcast said that there was a risk to humans because cow brains were going into the food chain. Articles in *The Guardian* and *The Times* began to suggest that meat pies and sausages posed a TSE risk, because the types of cow parts going into those foods weren't tracked. Worries about the public perception of meat safety led MAFF to issue a ban on the use of Specified Bovine Offal (SBO) for human consumption on November 13, 1989. Brains, spinal cord, and other organs deemed most likely to be infective could not be used in human food. It became known as the human SBO ban.

Britons weren't the only ones worried about mad cows. On July 28, 1989, the European Union banned U.K. cattle born after July 18, 1988. It was the first of many subsequent import restrictions that began that year by European countries and other nations, including the U.S.

Meanwhile, cases of BSE were skyrocketing. By the end of 1988, there were 2180 confirmed cases.[25] In 1989, 10,091 cows had come down with BSE; the number climbed to 24,396 in 1990. In the meantime, the prevailing thought among British officials was that the chance of a person contracting mad cow disease was so remote that it could be ignored so long as the human SBO ban was in effect. A Siamese cat, however, would rattle that complacency.

Mad Max

In December 1989, Jacqueline Stone of Bristol, in Avon County, southwest England, noticed something wrong with her five-year-old neutered male Siamese. While standing, Max would nod his head from side to side, as if he were about to doze off. By January 1990, Max had begun to stagger, the ataxia primarily affecting his hind legs. He became unusually startled by sudden noises. Stroking him down the spine evoked bouts of frantic licking and chewing at the stimulated area. His head soon adopted a slight tilt to the side. Veterinary work-ups didn't reveal any particular disease, and treatments with antibiotics and anti-inflammatories had no effect. The neurological symptoms progressed, and Max was euthanized in April 1990. Examination under the microscope revealed that parts of his brain had been eaten away, leaving the characteristic holes of a spongiform encephalopathy. Cats had contracted TSEs in the laboratory, but only when their brains were inoculated with infected tissue. Max evidently was the first cat to get the disease naturally, away from the syringe.

News of Max's death reached British ministers on May 6, 1990, and four days later, John Gummer, the head of MAFF, and David Maclean, the Parliamentary Secretary, met to decide how best to make the information public. In commenting to the news media, MAFF officials played down the significance of Max's diagnosis, stating that, although the investigation was continuing, there was no known connection between BSE and the cat, and the human risk remained the same.

Even so, the Siamese made front-page headlines, with one tabloid dubbing him "Mad Max." Stories made it clear that Max, who dined on cooked beef from the pet shop, had probably contracted the disease from his supper.[26] Local school authorities began pulling beef off the cafeteria menus. Richard Lacey, the microbiologist from Leeds who had criticized the government's handling of food-safety issues, stated that 6 million cattle—half of England's herds—should be culled as a precaution.

The government responded vigorously, issuing statements and press releases certifying the safety of British beef. One statement even suggested, erroneously, that one would have to eat impossibly large quantities of brain and spinal cord to be at risk. A newspaper's challenge to John Gummer to demonstrate that beef was safe led to a disastrous photo inopportunity. On May 16, BBC Newsnight showed Gummer trying to coax his hesitant four-year-old daughter Cordelia into biting into a burger. (The burger was reportedly too hot for her.) He later said he had done so because he had been assured that BSE posed no hazard—the human SBO ban, after all, was already in force.

Scientifically, Max was just one data point and might have been an anomaly. There was no proof that Max got sick from eating BSE-tainted food. Several zoo animals—exotics from the bovine family, such as the nyala, gemsbok, kudu, and eland, and from the feline family, including the tiger, puma, and cheetah—also got sick starting in the 1980s, all presumably from tainted feed. Yet the vast majority of animals, fed the same food, did not fall ill. But in terms of public health, Max raised the specter of a new black plague that had to be confronted.

So, given the rising public fears, scientists and government officials began reevaluating the risk. On May 16, 1990, the agricultural committee of the House of Commons began an investigation into BSE. On July 18, they reported that, although there was much uncertainty, there was no evidence of a risk to humans, considering that the source of infection had presumably been cut off. The Spongiform Encephalopathy Advisory Committee (SEAC), formed shortly after the Southwood Working Party and consisting of TSE researchers charged with providing advice to administrators, drew a similar conclusion. To the experts, the human SBO ban should make the risk to people remote.

Beef is safe, U.K. Minister of Agriculture John Gummer tried to show with his daughter Cordelia in May 1990. (*Jim James/PA Photos.*)

It would take other feline cases, however, to settle the question. In describing five infected cats in 1991, Wilesmith, Wells, and their colleagues wrote:

> The striking clinical and histopathological similarity of this disease in cats to the previously described transmissible spongiform encephalopathies of animals, leaves little doubt about the nature of the disease. The relatively widespread geographical distribution and pattern of emergence of these cases is similar to that witnessed at the start of the BSE epidemic, possibly because the disease in cats may have resulted from their exposure to a common source of infection.[27]

By December 1997, 84 house cats were reported to have contracted feline spongiform encephalopathy, or FSE. The number of actual FSE cases was probably much higher, especially in the late 1980s to early 1990s, before pet food makers stopped using offal in 1989 and the government banned SBO in all animal feed.[28] (Beginning in 1994, FSE cases had to be reported to the government.) The discovery of such cases, plus the deaths of various exotic zoo animals, offered solid evidence that the prion strain of BSE could jump to different species. Moreover, although lab experiments demonstrated that cats could contract a TSE, felines seem to be immune to scrapie. So whatever malformed prion protein was destroying their brains, it wasn't the same as the scrapie agent. If cows did indeed develop BSE because of scrapie, then the malformed prion protein must have taken a slightly new form that was hazardous to other species.

The question was whether this altered prion protein was transmissible to humans—could people succumb to a prion disease from mad cows? If so, it would probably resemble Creutzfeldt-Jakob disease, the most common prion disease of humans. But CJD produces a variety of different symptoms—how would this new strain from cows manifest itself? The job of looking for a new type of CJD fell to a tall, lanky neurologist from Edinburgh named Robert G. Will.

The Watcher

Western General Hospital is a sprawling complex on Crewe Road, lying about a mile northwest of Edinburgh's busy Princes Street and inconveniently obscured by a key on the official tourist map of the city. Near car park D next to the infectious diseases unit is a nondescript, one-story brick building, marked on the outside with a sign: CJD Surveillance Unit. This is the focal point for human prion diseases in the U.K., and Robert Will has been here since its founding more than a decade ago.[29]

"I'm a neurologist," Will explained. "As part of my training, in 1979 I applied for a job to do a search on CJD with Brian Matthews, a professor of neurology at Oxford. We did the first systematic epidemiological study in England and Wales."[30] That task gave Will the needed expertise to carry out one of the 1989 recommendations of the Southwood committee—namely, to monitor the prevalence of CJD in the U.K. to see if its epidemiology changed as a result of BSE. By that time, Matthews had retired, leaving Will as the one of the few people qualified for the assignment.

"We started in a porter cabin up the road there, just three of us," Will said. Neurologists would call the group when a suspicious case came up, and one of the researchers would go investigate. Eventually, the unit began incorporating neuropathological studies, under James Ironsides.

Today, the unit employs about 40 people and gets funding from various grants. The Department of Health has budgeted £1.2 million annually (about $1.9 million) for the unit until 2010, according to Will, who has been a long-time member of the Spongiform Encephalopathy Advisory Committee (SEAC).

Almost from the start, local neurologists referred twice as many cases to the surveillance unit than were actual CJD cases. "That was very important, because we couldn't predict what BSE would look like in humans, and finding everything that could be a spongiform encephalopathy in humans turned out to be particularly important," Will stated. Once the unit learns of a suspected case, it sends out a researcher to visit the family and to discuss the patient, who usually is in no position to respond by the time the CJD Surveillance Unit is involved. The visit includes questions about patients' medical histories, their diets, and how they liked their beef cooked. Families have reportedly been upset by some of the questions, such as whether the victim ate pet food or if she or he had used bone meal, dried blood, or manure.

"The expectation of finding something linked to BSE was not considered great, either by me or by the funders. I don't think anyone really expected we would find anything," Will recalled. "The general consensus was that BSE was unlikely to be a risk to the human population. That's what they said—everyone, including me. I didn't think it was very likely because of the evidence from scrapie." Will elaborated: "People forget, prion diseases in animals had *never* caused any problems in humans as far as we knew, despite extensive exposure."

Will and other scientists did know that the prion protein could change its properties when passing through a new host, "and that was the rationale for introducing all the public health measures in 1989 and 1990," Will said. "My view at that stage was if these measures were introduced and were enforced, which is what we believed, then the risk to the human population from BSE would be reduced enormously." In the minds of many, the SBO ban in human and animal foods went a long way to diminish any potential threat, so much so that "if you asked a group of scientists in 1990 what they felt the risks to the human population were, nearly all of them would have said the risks are remote."

The emergence of worrisome cases made the public increasingly anxious. One of the earliest publicized victims was Vicky Rimmer. Early in the summer of 1993, the 15-year-old from Connays Quay in North Wales began suffering from depression, memory lapses, and jerking movements, representing a neurodegenerative condition that her doctors could not diagnose. In September 1993, she went blind and slipped into a coma. Her grandmother claimed that Vicky became ill with mad cow disease because she ate beef. Researchers Stephen Dealler and Richard Lacey, both of whom frequently wrote that BSE presented more of a threat than government and other mainstream scientists were admitting, were cited as saying Vicky was the first victim of BSE, a highly speculative claim at that point. At the time, the CJD Surveillance Unit was uncertain if the disease was CJD.

Then there were the four CJD cases among dairy farmers between 1992 and 1995. Each had seen BSE in their cattle, and speculation ran rampant that they contracted the human form of mad cow disease. SEAC held one of its many special sessions in the fall of 1995 to consider the cases. This many farmers in such a short span may seem like a statistical anomaly, but the committee found that the blip was consistent with CJD in farmers in other European countries without BSE. More convincing were postmortems on the stricken farmers, which revealed neural damage no different from that caused by the classic, sporadic form of CJD.

Approaching the Watershed

Meanwhile, as Dave and Dot Churchill watched their son Stephen waste away in early 1995, they began making inquiries about CJD—the three letters they had seen on Stephen's folder at the National Hospital in London. They bumped into a couple in Devizes whose daughter had developed a negative reaction to a standard childhood vaccine. This conversation led the Churchills to another couple, who in turn referred them to a woman whose mother had died of CJD. That led them to Harash Narang, a microbiologist on the fringes of TSE research who was convinced that he had a test for BSE and was busy visiting families and victims. (His test, based on the idea that prion diseases are caused by an unusual virus he termed a nemavirus, could not be confirmed by other scientists.) "Narang was giving us information," Dave said. "We were starting to ask a lot more questions, to which we were getting no answers."[31]

Producers of *World in Action*, an investigative news television program in the U.K., wanted the Churchills' story. In August 1995, cameras followed Dave and Dot around as they gathered information and met with Richard Lacey, who had become prominent because of his BSE warnings. Journalists helped the Churchills along not just with the column inches and airtime, but with loads of information. "Some of them are so knowledgeable," Dot said. "They'd been following it since the '80s."[32] When the program and accompanying newspaper article ran on August 14, 1995, "there was an immediate denial in the same day" by the health authorities, Dave said, recalling that the government said that nothing had changed because of his son's death.

But things *were* changing. Peter Hall, a university student from Durham, began exhibiting symptoms when he was 19 years old and died 13 months later, in February 1996. Although vegetarian, he had loved to eat hamburgers as a child. In the fall of 1994, Christopher Pearson of Canterbury began noticing that his wife, Anna Pearson, a 29-year-old lawyer, was becoming forgetful; by March 1995, she had become so anxiety-ridden that she was terrified at the thought of going to court. She died one year later. Michelle Bowen, also 29, was pregnant with her third child when she developed symptoms early in 1995. She had to give birth three months prematurely via cesarean section and died later that year.

These and other cases began to worry Robert Will. Between 1970 and 1989, no one under 30 in the U.K. had contracted CJD; now, between 1990 and 1995, there were four definite cases and one possible case. It wasn't that young people were contracting CJD *per se* that concerned Will—it was the fact that the cases happened over a short period and that they were displaying a novel pathology. SEAC member John Collinge expressed his view that these cases likely represented BSE transmission to humans. Statistically, the cases were significant.

Complicating matters was the realization in 1995 that the human Specified Bovine Offal ban initiated in 1989 wasn't keeping risky material—brain and spinal cord particularly—out of the kitchen. The high-risk material probably entered sausages, meat pies, and burgers in the form of mechanically recovered meat. After most of the flesh is cut from the cattle carcass, the bones and other pieces are tossed into a machine. A piston in the device squeezes out the remaining bits of meat through sieves, resulting in a slurry that eventually becomes food. Originally, the SEAC concluded that such meat recovery wouldn't pose a problem if all the spinal cord were removed.

The problem was that not all spinal cord was being removed. In its meeting on November 28, 1995, SEAC learned that there had been 14 instances involving at least 25 carcasses in which high-risk material had been left on and could have gotten into the mechanically recovered meat. After a long debate, arguments by Will and the new committee chair, Sir John Pattison, convinced SEAC members that it was best to suspend the use of vertebrae from cattle older than six months in the production of mechanically recovered meat. The government implemented the ban in December 1995.

It turned out to be the right decision for a then-unknown reason: Clusters of nerve cells along the spine, called the dorsal root ganglia, had been making their way into mechanically recovered meat. These ganglia, which are junction boxes connecting the autonomic and the central nervous systems, were originally thought not to be infective and were thus exempt from the human SBO ban. Later research, however, showed that the dorsal root ganglia did indeed develop high levels of infectivity 32 to 40 months after a cow is infected with BSE.

By the end of winter in 1996, more evidence had emerged that a novel subset of CJD was striking young people. The CJD Surveillance Unit counted ten cases since 1994 (including Vicky Rimmer), and eight of the victims had died by early 1996. Will's colleague, neuropathologist James Ironsides, told the SEAC during its March 8, 1996, meeting that the young victims tended to have drawn-out symptoms. Another clue that a new disease had emerged lay in the neuropathology. In Stephen Churchill's brain, neurons had died off and spongy holes had appeared. Special star-shaped cells called astrocytes (under the microscope they are reminiscent of the black skate egg cases that wash up on beaches) had proliferated—a sign of the brain attempting damage control. The changes took place especially deep, just above Stephen's brainstem. Specifically, the portions involved were the hippocampus (critical to storing, sorting, and forming new memories), the thalamus (which relays sensory signals from the brain stem to areas of the cerebral cortex), and the basal ganglia, a collection of neuronal areas, including the putamen and caudate nucleus, that help control and coordinate movement. Stephen's cerebral cortex was largely spared of spongiform change. This 1/8-inch-thick gray outer layer of the brain controls higher mental functions such as language and conscious thought. The limited damage suggests that the speech impairment Stephen suffered late in the course of his disease was rooted in muscle coordination and not damage to his language center—he knew what to say but couldn't coordinate his larynx, lips, and tongue to form the words.

Although in prion diseases the parts of the brain attacked and the extent of the damage tend to be variable, in Stephen's case the areas were consistent with CJD's known targets. (The prime target in fatal insomnia is the thalamus; in kuru and Gerstmann-Sträussler-Scheinker syndrome, it is the cerebellum, which governs balance and movement.) What was puzzling in Stephen and other young victims were the "floral" plaques around the holes formed in the brain, particularly within the cerebellum. That had never been seen before, Ironsides told the SEAC.

The SEAC secretariat sent a memo to Sir Kenneth Calman, the U.K.'s chief medical officer until 1998, that told of SEAC's conclusion that exposure to BSE explained the novel CJD cases. Frantic memos and meetings between agricultural and health officials ensued, as authorities recognized that a national crisis was setting in. At an emergency SEAC meeting held on Saturday, March 16, 1996, Will presented data on nine confirmed and three suspected cases of the new variant of CJD that was affecting young people. Independent pathologists agreed

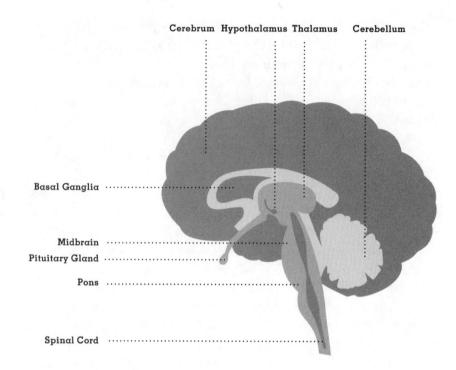

The human brain, shown as a vertical cross-section.

that the cases were distinct from previous CJD cases. The story leaked to newspapers that the government was about to announce that BSE could infect humans. On the afternoon of March 20, Stephen Dorrell, the Department of Health parliamentary undersecretary, stood before the House of Commons to relay SEAC's conclusions that the young victims' illnesses were linked to BSE.

"The world changed that day for us," Dot Churchill said. "We weren't the cranks," and neither were their supporters. Considering the consistent official stance that BSE was no real threat, the news was a bombshell to the public, which now felt its government had betrayed them. "March 20 is almost the watershed that September 11 is for the U.S.," Dave remarked. "It changed Britain to a large extent, changed eating habits, brought new legislation," and helped to drive the conservative Tory government out of office in the parliamentary elections of 1997.

The disease is called variant Creutzfeldt-Jakob disease. (It was called "new variant" in the April 6, 1996, *Lancet* write-up by Will, Ironsides, and their colleagues, although the "new" has since been dropped.) Although postmortems are the only definitive way to identify the disease, neurologists today can usually diagnose the condition in living patients based on the clinical symptoms. Unlike classic CJD, the variant form (abbreviated vCJD) starts out with psychiatric signs, affects young people, and progresses more slowly. EEGs and MRI scans of the brain can sometimes provide supporting evidence.

The name of the disease, however, still confuses both the public and some in the media, who forget to add "variant" to CJD. Other names have been floated, including "Will-Ironsides syndrome" by the National Institutes of Health's Paul Brown , without success. To the Churchills, their son Stephen died of BSE. If you contracted anthrax naturally, from a wild animal, Dave said, "people would say you died of anthrax." There's no formal way to name diseases. In the past, the diseases took on their discoverer's name, but that tradition has largely evaporated—understandably so for fatal conditions. A good name for a disease would actually say something about the condition; at least the name variant CJD puts it squarely in the realm of human prion diseases. "The media find it impossible to understand there are two forms of CJD," Will said, "and therefore they mix them up all the time and swap terms and cause confusion. And that upsets the families. I do regret that enormously. But it's the way it happened. And I don't think it can be changed now."

With the tide turned, influential scientists, members of Parliament, and the media, among others, began supporting the Churchills and other families of the vCJD victims in their quest for a public inquiry—an idea, Dot said, that originally came from U.K. reporter Alan Watkins. The Churchills had established the Human BSE Foundation and had been running a help line for vCJD families out of their home. "Initially, it was very hard bringing people together. But once they saw a focus, you couldn't keep them out," Dave recalled.

Three days before Christmas 1997, officials at MAFF and the Department of Health announced that a public inquiry would begin. It would be chaired by Lord Phillips of Worth Matravers. The BSE Inquiry began in January 1998 and was supposed to be done by the end of the year. But deluged with 3000 files and written statements from 630 witnesses, not to mention 138 hearing days for 333 witnesses, the inquiry dragged on for more than two-and-a-half years. The government released the report on October 26, 2000—a massive, 16-volume affair, taking up some 4000 pages that, in its less hefty digital form, still occupies more than 100 megabytes of disk space.[33] And that doesn't include all the supporting documents, transcripts of the hearings, and various written statements.

The report laid blame on several scientists and politicians for permitting the spread. Wilesmith's theory that BSE derived from scrapie and therefore might behave like the sheep disease in not being transmissible to humans blinded many researchers, even though they knew that the scrapie agent could adopt more infectious properties after passing to a fresh host. The scientists had offered various caveats—the Southwood committee acknowledged that if its assumptions were wrong, the implications would be extremely serious—but most of these statements were buried, the report concluded. When asked about the safety of beef, ministers and agricultural officials had recounted the committee's conclusions without conveying the uncertainty. Such misleading reassurances eroded public confidence in the government.

Lord Phillips, along with BSE Inquiry members June Bridgeman and Malcolm Ferguson-Smith (the lone scientist among them), also offered several lessons, including the need for greater public awareness about scientific uncertainty and the use of outside researchers. The MAFF had failed to deliver data to independent researchers, including epidemiologist Roy Anderson, who said that if a mathematical analysis were done in 1991, it could have shown that the ruminant feed ban of 1988 was in fact ineffective.[34]

The Inquiry produced reasonably fair conclusions, Will thought, "although it's terribly difficult in a situation like this not to use hindsight. I think it's unavoidable that you use the advantage of what you know now in judging what happened then." Although the report did point fingers, it did not produce the scapegoats that some may have hoped. "I don't think there was anyone doing anything deliberately bad," Will remarked. "I think people were trying to make decisions with the best available evidence at the time."

One item that the BSE Inquiry did not address was the media's influence on what scientists say. Will often found that his opinions and statements were presented in deceptive ways. In talking to a reporter when the CJD Surveillance Unit just got started, Will told him that "We have to continue this for many years because, if BSE is a risk, which we don't believe it is, it might have a long incubation period." Will then found out that "The headline of the paper was, 'Doctor warns about BSE time bomb.' That was not really what I was saying." Will added: "Some newspapers were quoting me when I hadn't even talked to them. They were making things up." Will discovered that this sort of news-spinning affects how you present the science and how willing you are to do so in the future.

A more contentious part of the report, however, was the Inquiry's insistence that BSE arose not from an endemic strain of scrapie in sheep, but from a random mutation in the prion gene of a cow or a sheep, much the way CJD spontaneously appears in one out of a million people. Hence, the Inquiry committee speculated, there were probably several undetected waves of BSE in the U.K. prior to its discovery in 1985. Many scientists prefer the endemic scrapie origin—it better explains why BSE showed up in England, which has a fairly high incidence of scrapie and has sheep as a greater fraction of rendered food material than other nations rearing both cattle and sheep.

To the families of the victims of variant Creutzfeldt-Jakob disease, the Inquiry did more than lay out the issues. It helped get them the care package that Dave and Dot Churchill had been fighting for. Because vCJD tends to strike the young, it often robs families of a wage earner

Mr. and Mrs. "Average" Britain: Dorothy ("Dot") and Dave Churchill, in their living room in Devizes, October 2001. (*Philip Yam.*)

or parent. The financial cost of caring for the sick varies—the Department of Health estimated it to be about £20,000 annually, while the lawyers for the families, taking into consideration time of care-givers, argued that it took £39,500 to £45,500 each year. After the BSE Inquiry was completed, the British government established a £1 million national care fund to provide better health and social support to vCJD patients. In 2001, the government made interim payments of £25,000 (nearly $40,000) to each vCJD family. If Stephen had gotten sick now, the Churchills would have been able to build a suitable bathroom in the house rather than leave him at the psychiatric ward or nursing home. And they would have gotten the wheelchair they had requested sooner, rather than having it arrive two weeks after Stephen's death.

The Churchills stepped down from the Human BSE Foundation after the end of the public hearings of the BSE Inquiry. "It was such a big part of our life," Dot said. "Once the inquiry was finished, we said it would be the start of a new era"—one in which the government is supposed to be more open and in which there are ways to address the special needs of vCJD patients. That's a sizeable accomplishment for the self-described "Mr. and Mrs. Average Britain with two children and a dog."

For the medical community, though, there was still the question of just how many families would have to go through the vCJD agony.

CHAPTER 9

Mad Cow's Human Toll

Figuring out how many people will succumb to variant Creutzfeldt-Jakob disease isn't easy—especially now that BSE has spread around the world.

For cows in the U.K., the worst is over. At its height, bovine spongiform encephalopathy was felling 1000 animals a week, in January 1993. It had hit its annual peak in 1992, with 36,682 confirmed cases. These cattle got infected before the ruminant feed ban of 1988, and because the average incubation period is five years (ranging from two to eight), numbers began dropping in 1993. They fell to 14,302 in 1995.

It quickly became apparent, however, that the 1988 ban wasn't perfect—some 36,000 cows born after the ban (called BABs) contracted BSE. Cross contamination was a problem; the ban did not prevent farmers from feeding ruminant protein to pigs and chickens, so some of this feed probably went to cows on mixed livestock farms.

On March 29, 1996, U.K. authorities amended the feed regulations to keep any mammalian meat-and-bone meal (MBM) from being fed to any livestock, fish, or horses. June saw the beginning of a massive recall of feed from all farms, mills, and merchants; some 10,000 metric tons were collected by October. In fact, August 1, 1996, introduced the "real" ban, as some observers called the regulation. At that point, it became illegal even to keep mammalian meat-and-bone meal with other livestock feed. Compliance has been high: Of 67,063 feed samples taken over about five years since February 1996, 99.74 percent tested negative for mammalian protein.[1]

These and other bans greatly reduced the BSE incidence even more, to 1202 cases in 2001 and about 1000 in 2002. Of lingering concern are the nearly two dozen BARBs — BSE cows born after the real ban of August 1, 1996. Low-level cross-contamination probably caused these cases (counted through August 30, 2002), although researchers have yet to rule out a vertical spread of BSE — that is, transmission from mother to calf. If such maternal spread is possible, it probably doesn't account for more than ten percent of all BSE cases, according to epidemiological calculations. Horizontal transmission between cattle — by direct contact between individual cows or with infected pastures — hasn't been ruled out definitively, but it seems highly unlikely. The European Commission's Scientific Steering Committee, a collection of European scientists who provide advice to governments, concluded in December 2001 that feed and maternal transmission were the only possible routes of spreading mad cow disease.

In total, by the end of 2002, the U.K. had seen nearly 180,000 BSE cases on almost 36,000 farms.[2] How many BSE cases escaped detection and went on to infect humans remains a matter of informed speculation dependent on several assumptions.

Calculating Mortality

The first and most basic assumption is that BSE really does cause variant Creutzfeldt-Jakob disease in humans. Of course, the only way to be sure is to inject BSE tissue into a live human brain and see if the characteristic floral plaques and holes appear. Although there's no proof, Robert Will of the CJD Surveillance Unit explained, "I think there's very powerful evidence, both epidemiological and scientific."[3] Rodents inoculated with brain tissue from humans who had vCJD, from cows with BSE, and from domestic cats with FSE all developed similar neural lesions. Molecular analyses also substantiated the link: The prion proteins from vCJD humans, BSE cows, FSE cats, and BSE-inoculated macaques were all broken down to the same extent after exposure to the digesting enzyme proteinase K. In other words, the proteinase K-resistant cores of PrP^{Sc} from all the species were similarly sized, suggesting that all those animals had the same strain of spongiform encephalopathy. "They all add up in a sort of series of steps that suggest quite powerfully that BSE is the cause of vCJD," Will elaborated. "What we don't know is, how the transmission occurred"—whether, say, the BSE prions are absorbed by the gut after a meal or invade through a cut or sore in the mouth. "That's the biggest weakness in the theory," Will stated.

A second assumption needed to estimate the eventual vCJD toll is the amount of infected beef that might have made it to the dinner table. Or put another way, how many BSE cows were actually staggering around Britain? Because BSE has a long incubation period, averaging five years, many subclinical cases—infected cows that seem healthy—probably entered the food chain. In October 2002, epidemiologists Roy M. Anderson, Christi A. Donnelly, and their colleagues at Imperial College in London reported an estimate derived from a statistical technique called back-calculation, a method used successfully to determine the scope of the HIV/AIDS epidemic in various populations. The method relies on current case data to determine the conditions that originally sparked the epidemic. From that, researchers can work backward to estimate the number of individuals that became infected.

Their "differential slaughter model" assumes that early cases went unnoticed. It also supposes farmers may have inadvertently slaughtered subclinical cows preferentially, because the early stages of undiagnosed BSE may have decreased the milk output of a cow and thereby encouraged a farmer to send the animal to slaughter. With those assumptions, the epidemiologists concluded that between 1980 and 1996, 1.9 million cattle were affected by BSE and 1.6 million subclinical cases became food for humans. That's a near doubling of the team's 1996 estimate of 1 million infected cattle and 750,000 asymptomatic BSE cows slaughtered for human consumption.[4] The bump-up is actually good news, because it suggests that sick cows are less infective than originally thought—it took more BSE cases than previously thought to produce the current vCJD numbers.

In 1999, the European Commission's Scientific Steering Committee suggested that each cow could have exposed 500,000 people to BSE. (The estimate assumes any product derived from a cow, regardless of the amount used in the product.) So with 1.6 million infected cattle having presumably entered the food chain, potentially 800 billion doses of BSE were consumed between 1980 and 1996. That would imply that virtually all of the 60 million people living in the U.K. before 1996, and any visitors, were exposed to a massive amount of infected material. Not all the English have gone mad and died, meaning that several other factors must be at play in determining who succumbs to vCJD.

The various control measures certainly cut down the potential exposure. The first, the Specified Bovine Offal (SBO) ban of 1989, prevented a lot of high-risk bovine material from entering human food, but certainly not all of it. In June 1994, researchers learned that additional parts of the cow were also risky. This knowledge came to light during the course of an experiment at the Central Veterinary Lab. The study involved feeding BSE-infected brain to calves and then slaughtering one every four months. Samples were collected from 44 different tissues in the calf's body and inoculated into the brains of mice to determine which parts of the calf had become infected and when. In calves under six months of age, the distal ileum, part of the small intestine, proved to be infective. That, plus the thymus, tonsils, and other parts of the lymphoreticular system were soon banned. After December 1995, slaughterhouses were not allowed to recover meat mechanically from the vertebrae, because it often has bits of highly risky spinal cord and dorsal root ganglia still attached.

Once BSE moved from being a theoretical risk to humans to a probable one in March 1996, the government imposed additional measures to provide an extra margin of safety. All cattle older than 30 months were banned from human food entirely. (Although most beef cattle are slaughtered when they are younger than 24 months anyway, the ban did prevent dairy cattle at the end of their milking life from being sold as food.) The whole head of the cow except for the tongue was also banned for fear of contamination from the brain during butchering. In December 1997, the U.K. excluded the sale of beef on the bone, including T-bone steaks and ribs. (The ban was largely lifted by the end of 1999.) In January 1997, the government began a program to cull any animal that might have been exposed to infected feed.

Although of concern at the start of the BSE outbreak, milk has not been shown to be infective or to harbor prion protein. In 1995, David Taylor and his colleagues at the Neuropathogenesis Unit in Edinburgh reported on experiments in which mice were inoculated or were forced to drink large volumes of milk from BSE cows; the rodents remained healthy. Restrictions that prevented milk, as well as gelatin and certain blood components, from being used in feed was lifted on March 6, 1995.

Existing vCJD data is also critical in determining the future course of the epidemic. As of March 2003, the CJD Surveillance Unit had logged 132 definite and probable cases of vCJD in the U.K. The course of the illness ranges from 6 to 39 months, the average duration being 13 months. Psychiatric symptoms such as depression, anxiety, insomnia, and irritability tend to dominate at first. Clumsiness in walking and slurring of speech follow 4 to 6 months later. About 15 percent of cases present neurological symptoms—headaches, muscle weakness, loss of consciousness—before the psychiatric signs.[5]

No one really knows why vCJD attacks some people but not others, particularly among family members, who presumably have eaten similar foods for years. (The only variant CJD cluster was five cases in the small town of Queniborough, Leicestershire; three died within 12 weeks of one another. There, traditional butchering with a contaminated knife may have caused the outbreak.) Variant CJD does prefer the young. Of the first 100 victims, the average age was 29, with the youngest being only 12. (The oldest victim to succumb was 74 years old.) One speculation is that younger people tend to eat more prepared meat products, such as sausages and pasties, which likely incorporated the high-risk mechanically recovered meat. To neuropathologist Stephen DeArmond of the University of California in San Francisco, that theory doesn't wash. "I go over to England and see a bunch of old-timers like myself sitting around in the pubs, eating bangers and mash, and meat pies. Seems like an awful lot of people eat what we would call junk food. I even do it."[6]

Rather, a clue may lie in the way the disease first shows up in the body. The rogue prions appear to attack the tonsils and lymph nodes first, before making their way to the brain. "My wife has always told me—she was a pediatrician—that the tonsils are still active in young people and begin to regress in your teenage years and into the 20s," DeArmond said. People can pick up infections from many common sources—a sibling's sneeze, say, or grimy fingernails handling food. "Theoretically, we have at least two viral infections at least a year. After 50 years, you've had 100 different viruses that you become immune to. But a young person still has to go through all that. Many of those viruses cause sore throats and activate the tonsils and lymphoid tissue." Such activation causes the tonsils to swell and become inflamed, presumably making it easier for BSE prions to enter at that point—in other words, having a cold could have predisposed young people to contracting vCJD.

The number of sick cows, the control measures used to halt the spread of BSE, and the disease profile all figure in modeling the number of future vCJD deaths. With these and other factors, and working with various assumptions, Anderson and his Imperial College colleagues ended up with a highly complex model with many flexible parameters. In all epidemiological models, the most critical parameter is the incubation time—how long it takes to come down with symptoms after you have been infected.

Stephen Churchill and the others who died in 1995 represent the first known instances of BSE infecting humans. Optimists assume that that the riskiest time occurred when the BSE epidemic was running high, before the 1989 human SBO ban. If this is true, then the incubation period of vCJD would be about five to six years and the worst would be over and the number of vCJD cases should be dropping. The more pessimistic view assumes that the first vCJD infections could have occurred when cows first became infected, around 1980. That assumption suggests an incubation period of 10 to 15 years, more in line with what happens when prion diseases cross species—that is, the incubation period lengthens for a new host species. In that case, the peak of the vCJD epidemic should fall between 1999 and 2004.

Assuming an incubation period of less than 20 years, the U.K. will probably experience only several hundred vCJD cases, according to an August 2000 estimate by Anderson and his colleagues. Their estimate for the maximum number of cases was 136,000 —and that number assumes an incubation period that can be as long as 60 years. "This would make it unusual, but it cannot be ruled out," remarked Anderson's colleague Neil M. Ferguson, especially considering that kuru can incubate for more than 40 years.[7] In 2002, the team lowered its estimates, giving a range of 50 to 50,000 vCJD deaths between 2001 and 2080; in February 2003, it dropped its estimate further, to 10 to 7,000 deaths.[8]

The February calculation is in line with the 2001 findings of another research group, led by Peter G. Smith of the London School of Hygiene and Tropical Medicine. It relied on back-calculation, extrapolating from current vCJD cases. Only seven variable parameters entered the equation. They concluded that there would be no more than 10,000 cases and, more likely, only a few hundred. The worst-case scenario, in which millions are infected, would mean that the incubation period is so long that most people will die from other causes before developing clinical signs of the disease.[9]

Although the research teams have proposed different estimates, both sets of projections appear to convey good news. Early on, when the very first cases were coming in, estimates of future vCJD deaths ranged into the millions. Still, "all these estimates are very flexible and fragile, related to various assumptions," Robert Will said. One is age: The discovery of a 74-year-old with vCJD "immediately increased the number significantly," Will noted. "I have to say myself that I think it's very difficult to rely firmly on mathematical calculations when you're at the start" of a potential epidemic of a disease with very long incubation periods. "I think it's difficult to be confident that the predictions are accurate at the present time when there are so many uncertainties."

A complicating uncertainty is dose. Although a thimbleful of BSE tissue can bring down a cow, no one knows how much BSE meat is needed for humans to contract vCJD. Based on his team's calculations, Ferguson thinks the species barrier between humans and cattle is high and concludes that no more than two vCJD cases would emerge from a maximally infected BSE cow. (This figure represents a drop from an earlier estimate of 100 vCJD cases per animal because increased numbers of BSE cows and vCJD humans permitted more detailed calculations.) "The questions about the absolute risk of infection from eating particular tissues from an infected animal cannot be answered, as we have no idea what the infectious dose is for humans. If we could answer that question, we could estimate the potential scale of the vCJD epidemic much better than we can at the moment," Ferguson commented. In its travel advisory, the U.S. Centers for Disease Control and Prevention states that the current risk in the U.K. appears small, perhaps about 1 case per 10 billion servings of beef.

The estimates also focus on humans who have a particular genetic makeup. The appearance of vCJD depends on codon 129 of the PrP gene. In other prion diseases, the type of amino acid specified by codon 129 sets the incubation time and even the type of hereditary prion disease, dictating, for instance, whether one comes down with sporadic

CJD or fatal familial insomnia. In the vCJD epidemic so far, victims to date are methionine homozygous at codon 129 — that is, both their PrP genes (one from each parent) have the amino acid methionine (M) coded at position 129. About 37 percent of the Caucasian population are M/M at codon 129. The remaining 63 percent have two valines (V) or a methionine-valine combination at codon 129; no one with these genotypes has contracted vCJD yet.

The V/V or M/V genotypes may be immune to the disease. Or perhaps these people can just fend off the rogue proteins longer. If the latter is true, the ultimate number of casualties could jump by a factor of 2.5. Among methionine homozygous people, researchers assumed that there is no other genetic influence on the incubation period — that is, there is no subset of M/M individuals who are better at fighting off BSE prions and thus have extended incubation periods.

Mad Sheep Disease?

The estimates of vCJD cases also assume that the mode of infection was eating beef. But early on, sheep also ate contaminated meat-and-bone meal, and experiments reported in 1993 showed that sheep could get BSE when fed infected cattle brains. Did British sheep contract BSE — and did people contract vCJD from eating lamb rather than beef?

In 1997, Chris Bostock of the Institute for Animal Health in Compton, Berkshire, attempted to determine just that. His team took stored brain material pooled from 3000 sheep that were diagnosed with scrapie in 1992. The researchers injected bits of the sheep brain into mice genetically engineered to succumb to the BSE prion protein, a process that would take a couple of years. The mice started getting sick, which suggested that indeed the sheep did contract BSE and that between 0.1 percent and 1 percent of U.K. sheep were infected. Considering that scrapie transmitted horizontally, it was assumed that BSE in sheep might do so as well, so that just a 0.1-percent incidence would have serious implications for public health. The government even prepared a contingency plan to slaughter each and every one of the nation's sheep.

On October 17, 2001, however, the Department for Environment, Food, and Rural Affairs (DEFRA, which replaced the now defunct Ministry of Agriculture, Fisheries, and Food) announced that a terrible mistake had been made. In testing the samples inoculated into the mice, the Laboratory of the Government Chemist concluded that the brain tissue in the mouse bioassays had come from BSE cattle, not sheep. A mix-up of pooled samples had occurred, making the results of the four-year, £217,000 ($308,000) experiment worthless.[10] Two independent audits concluded that a labeling error occurred sometime in the mid-1990s among brain samples stored at the Neuropathogenesis Unit, the Edinburgh branch of the Institute for Animal Health.

If sheep did contract BSE in the early 1990s, the numbers were probably small—calculated to be between 10 and 1500 sheep in total. The annual incidence, assuming limited maternal transmission, would be fewer than 20 BSE sheep, hidden among 10,000 scrapie sheep. Still, sheep with BSE would mean higher vCJD numbers. The upper limit of vCJD cases calculated by Neil Ferguson and his fellow epidemiologists at Imperial College in London would rise by a factor of 3, largely because the stringent controls protecting human health from BSE cattle were not in place for sheep. Ferguson noted that two measures could greatly reduce the risk from sheep-borne BSE: banning the use of all offal in sheep feed and ensuring that animals are slaughtered when young (under 12 months), when they have less infectious prion protein in their bodies.

Spreading the Madness

In December 1989, Hilary Pickles, then a high-level medical officer for the Department of Health in the U.K., wrote a disturbing memo to the Chief Medical Officer, Sir Donald Acheson. It summed up her feelings and those of several colleagues about the then-emerging BSE crisis:

> This concerns the continued export of potentially BSE- and scrapie-contaminated meat and bone meal from the UK. We acted promptly in this country to ban the feeding of this material to ruminants last summer. The tardy response from other nations, with so far only one

or two restricting use of UK imports, suggests that the risk has not been fully appreciated overseas. Indeed it is unrealistic to expect nations who have not seen any BSE (yet) to give this any priority.[11]

That warning proved true for Ireland, which discovered its first case of BSE in 1989; Oman and the Falkland Islands also reported BSE cases that year, because they had imported subclinical BSE cows. In 1990, Portugal and Switzerland joined the mad-cow club, followed by France, Denmark, and Germany in 1992. Those cases resulted from either the importation of infected cattle or from the use of contaminated meat-and-bone meal shipped out of the U.K. before the 1988 regulations that prohibited the feeding of ruminants to ruminants. After the U.K. instituted that ban, feed makers apparently dumped their supplies onto the world markets: In 1988, the U.K. exported 12,533 metric tons of meat-and-bone meal; the next year, it exported 25,000 metric tons to the countries of the European Union and another 7,000 tons to nations of the Middle East and Africa. Most of Europe stopped taking U.K. meat-and-bone meal by 1991, but shipments to other areas persisted—the developing world took in some 30,000 metric tons that year.[12] After 1991, the amount exported has been in the range of a couple thousand tons each year.[13] The countries that imported the most during the height of the BSE epidemic were Indonesia, India, Thailand, Taiwan, and Sri Lanka. If they set up formal surveillance programs, they would likely find cases of BSE.

Indeed, the Food and Agriculture Organization of the United Nations decided that all countries that imported meat-and-bone meal made in the U.K. in the 1980s are potential incubators of BSE. That's more than 100 countries. (Many nations imported U.K. feed, processed it, and then re-exported it.) The justification for the export, which continued until 1996, was that the importing countries had a clear understanding that the feed must not be given to cattle—it was meant only for chicken and pigs. Yet Prosper de Mulder, Britain's largest processor, acknowledged that the feed could have been used for cattle.[14] Moreover, many countries continued to accept MBM from Europe, where more BSE cases were turning up. For instance, Japan only stopped importing the ruminant-based feed from continental Europe in 2001; moreover, it continued to feed domestically made mammalian protein to their cattle, thus potentially amplifying any lurking BSE.

The British government's handling of the BSE crises should have been a lesson to other nations in the world. So it was rather a surprise to see how poorly Japan dealt with the BSE risk, ignoring it early on. Tokyo agreed to join the European Union in a geographic risk assessment of BSE conducted by the European Commission's Scientific Steering Committee. But even as they participated in this assessment, Japan believed that it did not have the disease that by early 2001 was causing widespread fear among European consumers as more nations kept finding BSE cases. (The seemingly sudden surge of dozens to hundreds of cases per country resulted largely from active surveillance programs, in which cattle tissue is routinely tested for PrP^{Sc}.) In June 2001, after Takashi Onondera, chairman of Japan's agriculture ministry's BSE committee, learned of the Committee's upcoming opinion on the country's BSE risk, Tokyo pulled out of the study and stopped sending in data.[15] The Committee was to conclude that Japan was at high risk for having BSE and, because it made cattle feed from other (local) cattle, its feeding practice would spread the contagion.

As the Japanese government turned a blind eye, the risk assessment proved all too accurate. In early August 2001, a five-year-old Holstein from the city of Shiroi, east of Tokyo in the Chiba Prefecture, started coming down with BSE symptoms. Initially, veterinarians thought the cow's illness was septicemia stemming from a bacterial infection. The cow was slaughtered on August 6 and deemed unfit for human consumption. An immunoassay test on a brain tissue sample, meant to detect the presence of PrP^{Sc}, came back negative on August 15. But three other tests done during the next three weeks—two of which had pathologists look for the characteristic vacuoles of spongiform encephalopathy—were positive for BSE. On September 10, the Japanese Ministry of Agriculture, Forestry, and Fisheries alerted the *Office International des Epizooties* (OIE, the Paris-based World Organization for Animal Health) to their suspected BSE case and sent tissue samples to Britain and Sweden for confirmation.

Japanese officials quarantined the herd and stated that the carcass of the BSE cow had been incinerated. Four days later, they admitted that in fact the cow had been turned into feed and went into a total of 145 tons of meat-and-bone meal.[16] Japan instituted a cull of about 5000 cows that had been fed MBM, although it was far too late to keep the disease out of the country—two more BSE cases turned up by

November 30, 2001, another two in 2002, and yet another two in January 2003.

No one is really sure how Japan got BSE. Possible culprits are imported meat-and-bone meal or even scrapie sheep rendered into cattle feed. (Japanese farmers were not required to incinerate scrapie sheep until 1996.) Japanese officials focused on a substitute milk feed—made from skim milk powder and protein from pig and beef tallow.

Although the number of sick cows in 2001 was tiny when compared with the European numbers, Japanese consumer confidence in both their food and their leaders evaporated. Shortly after the news of the first BSE cows broke, nearly 2000 schools pulled beef off the menus. Supermarkets and restaurants such as McDonald's, which has 3700 outlets in the country, posted signs declaring that their beef came from Australia and the U.S., where BSE had not been detected. In a scene reminiscent of John Gummer's disastrous 1990 burger-eating photo session with his daughter, Japan's agriculture minister, Tsutomu Takebe, scarfed down some grilled steak before television cameras in October 2001, declaring he never tasted such good beef.[17] Widely criticized for his handling of the BSE crisis, Takebe barely held onto his job when the Japanese parliament rejected a no-confidence motion against him in February 2002.

Factoring in lost beef sales as well as compensation payments to farmers, the first three BSE cows are estimated to have cost the Japanese economy more than ¥365 billion (about $2.76 billion). There have been no vCJD deaths in Japan—the only human casualty so far was a veterinarian who committed suicide, reportedly because she felt guilty for having missed the BSE diagnosis. Considering that 92 percent of the Japanese population are methionine homozygous at codon 129 (compared with the 37 percent or so of the European population), a potential vCJD epidemic in Japan would be much worse than in the U.K. (By the end of 2002, vCJD had occurred six times in France, once in Ireland, and once in Italy. The lone U.S. case was a British transplant, and the only Canadian case was an individual who had spent several years in the U.K. during the BSE epidemic.)

As of December 2002, the OIE recorded 22 countries that have confirmed BSE in their herds. Ireland had the second-most cases after the U.K., having seen a total of 1150 by December 2002. Portugal and France had found about 700 BSE cases each, and Germany 235. Italy

Japanese health minister Chikara Sakaguchi, left, and agriculture minister Tsutomu Takebe take a page from John Gummer's book (see page 123) and show their
faith in the safety of beef in October 2001, shortly after Japan discovered its first
mad cow. Several other BSE cases soon turned up. (*Toshiyuki Aizawa / Reuters.*)

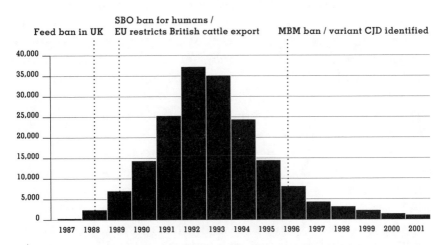

BSE cases in the U.K. peaked in 1992. Because of BSE's long incubation period, several
years pass before the effectiveness of various bans can be seen. (*Office International
des Epizooties / U.K. Department for Environment, Food and Rural Affairs.*)

had only four dozen cases in 2001. (Because the numbers of confirmed BSE cattle were so low, the one vCJD victim in Italy raised concern that BSE could jump to humans more readily than previously thought.) These cases probably stemmed from animal recycling after an initial infection of imported feed from the U.K., because the European Union did not ban the feeding of mammalian protein to ruminants until 1994—years of watching the British BSE experience apparently did not trigger any sense of urgency.

Other measures were introduced over the years, some to bring a negligible risk down even more. One ban introduced in the European Union beginning in 2001 was a prohibition of penetrative stunning methods in slaughterhouses. Before cows are bled to death, they are knocked unconscious by a "captive bolt" fired at their skulls (the bolt retracts into the gun, hence the term "captive"). Some guns drive the bolts so powerfully—they travel about 100 meters per second, or about 220 miles per hour—that they can blast through the skull; other types of guns even inject air into the skull to literally scramble the brains. Infusing air can send good size bits of brain down the spinal column and into the circulatory system, where they can end up in the heart, lungs, and even kidneys. Penetrative stunning could thus cause brain matter to contaminate cattle parts not thought to pose a BSE risk to humans. Presumably, if only cows under 30 months are being slaughtered (the standard now in Europe), then the brain isn't likely to contain many BSE prions, if any. In any case, slaughterhouses now mostly rely on non-penetrating captive bolts, which the Scientific Steering Committee considers to present no more of a contamination risk than kosher and halal slaughtering. These Jewish and Islamic practices call for exsanguination without stunning and probably offer the safest slaughtering method in terms of preventing brain tissue from reaching other areas of the cow.[18]

Strong regulations should contain the BSE outbreak—at least in principle. But the practice of those rules is sometimes spotty. Britain has had some cross-contamination, as revealed by a handful of BSE cows born after August 1, 1996, when stringent feed rules came into force. Germany's beef market was clobbered in 2001 after it had found 125 BSE cases that year—ten times more than it had found up until then. In early 2002, the government acknowledged that some 40,000 beef carcasses heading for human consumption were tested by an unauthorized laboratory in Bavaria. McDonald's, which has 1200 outlets in Germany, pulled some of their beef from circulation because another lab was found not to have properly tested it. In July 2002, the French food safety agency reported that in about 10 percent of the cattle carcasses about to become human food, slaughterhouses had left risky material such as spinal cord, tonsils, and thymus. Some pieces were as long as 8 inches. (A newly instituted vacuum system, rather than manual cutting, should strip carcasses more cleanly.)

Although crises pass, mad cow disease has left lingering aftereffects when it comes to economics. The height of Germany's outbreak took place in early 2001, which naturally drove people from a favorite meat to alternatives like chicken and fish. Sales of beef recovered somewhat by October 2001, but even a year later, beef sales lagged behind Germany's pre-BSE days. Countries have engaged in beef import bans at various times, leading to much ill will and retaliation. The U.K. bore the brunt of other nations' ire, and British beef understandably was prohibited from the continent for a few years. The European Union began accepting British beef in 1999, and even long-time holdout France finally lifted its six-year-old import ban in 2002. (France had been Britain's biggest beef importer until it stopped in 1996.)

BSE probably exists in many developing nations that took in a lot of U.K. feed in the 1990s. Many lack the means to establish a formal surveillance system and have more pressing problems, so it is impossible to gauge the true spread of mad cow disease. Several countries have asked for the opinions of the European Commission's Scientific Steering Committee about the chances of seeing BSE in their herds. The Committee assesses a country's mad cow risk based on its potential exposure to BSE and its ability to control an outbreak. In June 2000, the Committee completed its first round of geographic risk assessments for its member states and certain other countries; since then, it has assessed more than three dozen other nations. It slots countries into four possible categories. On one end are Category I countries, which are considered highly unlikely to have BSE cows: Argentina, Australia, Botswana, Brazil, Chile, Costa Rica, El Salvador, Namibia, Nicaragua, Norway, New Zealand, Panama, Paraguay, Singapore, Swaziland, and Uruguay. The other extreme is Category IV, in which BSE is confirmed at comparatively high levels; as of 2001, just the U.K. and Portugal fell into this category. Most of the rest of Europe, having found several cases of BSE, is classified as Category III, which also now includes several eastern European and Mediterranean nations. Category III also includes countries that are likely to have BSE, although they may not yet have confirmed it; Albania, Bulgaria, Croatia, Cyprus, and Turkey fall into this category, and there soon may be others.[19] (Japan would have fallen into Category III had it not pulled out of the assessment.)

The U.S. falls into Category II: Although an epidemic is unlikely, a BSE risk cannot be excluded.

CHAPTER 10

Keeping the Madness Out

Several measures help ensure that animal prion diseases do not contaminate the U.S. food supply—but there are gaps.

World travel and trade have brought numerous economic benefits to the U.S. but all too often have also delivered unwanted goods—from ecological destruction by zebra mussels and long-horned beetles to illnesses by deadly germs such as the West Nile virus. Did the U.S. also unwittingly import bovine spongiform encephalopathy?

Millions of pounds of beef and beef products from countries later found to have mad cow disease landed on American shores: Between 1980, when BSE probably began emerging, and 2000, the U.S. took in about 1000 cattle, some 50,500 tons of beef, 11,500 tons of meat by-products, including meat-and-bone meal, and 12,000 tons of prepared beef products. The U.S. imported 334 breeding and dairy cows from the U.K. between 1980 and 1989, and 173 of them could have been eaten by humans (or by other cows in the form of feed). The import numbers are small compared with overall figures—none of them amount to more than 0.73 percent of the total for their category (imported cattle being the least at 0.003 percent).[1] But as the BSE epidemic in the U.K. showed, it just takes a thimbleful to make a cow sick, and rendering and feed practices can transform a pinch of prions into a widespread disaster.

Cows in the Crosshairs

A name like Corporate Boulevard might conjure visions of sleek, steel-and-glass office buildings lining the street and rows upon rows of parked cars. One such boulevard in New Jersey, however, doesn't even have lanes painted on it. About an hour's drive down the New Jersey Turnpike from New York City, this dead-end road in Robbinsville, New Jersey, leads to a modest, one-story building complex that houses the offices of several organizations. In one corner is the Animal and Plant Health Inspection Service (APHIS), the arm of the U.S. Department of Agriculture (USDA), whose mission is to "protect American agriculture." In this particular APHIS field office is a curly-haired woman who seems like an unlikely target for angry phone calls, hate mail, and occasional death threats. Such is the life of Linda Detwiler, New Jersey farm girl turned government senior staff veterinarian who has been in charge of looking out for prion diseases in U.S. livestock since 1996.

Detwiler doesn't literally stand in pastures with a pair of binoculars, of course. She coordinates the APHIS surveillance, prevention, and education activities, has served on various working groups and TSE advisory committees in the U.S. and abroad, helped develop the U.S. response plan in the event a mad cow is discovered, and is the media spokesperson for TSE-related issues. Thrust in front of reporters, she had to explain why the USDA had moved in to "depopulate" two flocks of sheep in Vermont in March 2001, which were imported from Belgium and may have been exposed to contaminated feed. To those who found the action intrusive, she became "Dr. Deathwiler" and a target for frustration and anger.

She seems to take it all in stride. "I can remember one day in particular when everything I picked up had foul language," Detwiler recalled. "That's part of the job." With a chuckle, she paraphrased Nietzsche: "What won't hurt you makes you stronger."[2] She prides herself on returning serious calls from concerned citizens to explain USDA actions, and her dedication and expertise earns her high marks from TSE researchers around the nation. Some vindication of the Vermont sheep episode came her way in April 2002 when the USDA announced that 6 imported sheep out of 380 confiscated (the latter number included a flock voluntarily sold in July 2000) did in fact test positive for some type of TSE. To determine whether it was a strain of scrapie or BSE in sheep requires mouse bioassays, the results of which are expected by 2005.

The USDA plays two diametrically opposed roles: a consumer protector, ensuring the safety of the nation's food supply, and an advocate for industry, helping ranchers, farmers, and others in the supply chain sell their goods. These roles represent a conflict of interest, so it is natural to wonder whether mad cows might secretly be wandering on American feed lots. Just three BSE cases pummeled Japan with an estimated $2.76 billion in total costs—a number that factored in government compensation to cattlemen, lost farm revenue, slumping meat sales, and diner rejection of *bugolgi* (barbecued beef) at Korean restaurants.[3] Even if Japan's agriculture ministry overestimated the costs, a single case of BSE undoubtedly will cause economic headaches throughout many sectors. In 1998, the Food and Drug Administration concluded that if the U.S. had a mad-cow outbreak as bad as the U.K.'s, lost sales revenue alone would be upward of $15 billion, assuming a 24-percent drop in domestic beef sales and an 80-percent drop in exports. Then, factor in another $12 billion for slaughter and disposal costs of at-risk animals.[4] Considering the power of the $56-billion-a-year cattle industry, it just doesn't seem economically or politically expedient to find BSE. "It's almost a 'don't look, don't find'" attitude, remarked Michael Hansen of Consumers Union of Yonkers, New York, a longtime critic of the U.S. approach to handling TSEs. "You don't want to look too hard. If they find things, you'll have a short term economic calamity."[5]

The U.S didn't look very hard early on. It began testing cattle brains for BSE in 1990. But throughout the decade, only several hundred were tested each year—a mere 0.0007 percent of the total U.S. bovine population of about 100 million. Almost half the tested cows fell into the groups most likely to harbor BSE: They were neurologically ill or could not walk for whatever reason. These nonambulatory animals, usually called downers, are those cattle that cannot stand on their own at any point in a 24-hour period. Many ailments can cause this problem, including broken limbs, arthritis, mineral deficiencies—and brain damage. From fiscal years 1994 through 1999, the average annual number of downer brains tested in the U.S. was 317; for all cow brains, it was 687.

Such a low level of testing would have been unlikely to have spotted any mad cows even if the incidence were as high as was seen in France, the country with the most cases after the U.K. and Ireland. In 2001, its peak BSE year, France uncovered 191 BSE cases out of 134,358 tests on high-risk cows—those clinically suspected of having BSE and those that seemed sick at slaughter or died on farms. This ratio suggests that 704 cows must be tested to find 1 BSE case. The closest the U.S. came to this number in the 1990s was in 1999, when it tested 651 downers. U.S. testing numbers began improving substantially by the turn of the millennium, going to 2309 in 2000 and 5272 in 2001. It reached 19,900 in 2002—significantly more than the originally planned 12,500 for that fiscal year.

On the face of it, that number may still seem too low. In some European countries, BSE occurs at an annual rate of a few per million, suggesting that hundreds of thousands of cattle should be tested. Germany and France each tested more than 2.5 million cattle in 2001, the vast majority of which were healthy animals. More comparable to the U.S. situation is Austria, which is also a Category II country in terms of BSE risk. It screened some 225,000 cattle to find its first mad cow in 2001.[6]

As Detwiler put it, surveillance is more than a numbers game: "It depends on the population you're testing and how good your rate of return is." The U.S. approach has focused on downer cows, of which there are about 200,000 in the U.S. every year, according to Detwiler, "and if you test 12,500 [the original 2002 testing target] out of that population, you should be able to detect it at that rate of one per million." The difference between the U.S. and Europe is that "we're targeting the highest risk and not the normals going to slaughter," as European nations are. "They are not doing it to find BSE there. They're doing it to try to pull [BSE] animals out of the food chain."

Paul Brown, who has served with Detwiler on TSE advisory committees, agrees with the U.S.'s strategy. Null test results for neurologically ill and downer cows are "much more significant as a negative than it would be as a random" sampling of healthy cows, he noted. "A lot of the public doesn't understand that. They say, 'God, France is doing tens of thousands of tests,' and so forth. In France, that is perfectly legitimate—Christ, it's chock full of BSE," Brown observed, thanks to its importing tens of thousands of tons of meat-and-bone meal during the period of high risk and recycling infected tissue through rendering.[7] In fact, the U.S.'s 2002 testing numbers exceeded the standard for

Category II countries suggested by the OIE, the World Organization for Animal Health, by more than 40 times.

Detwiler draws an analogy: "Say we didn't know we had Alzheimer's in the U.S. and you had to set up a program to test for it. What would you suggest the population to be tested? Older population, maybe with signs of dementia, right?" Testing cattle at slaughter, which for 88 percent of cattle occurs at less than 18 months of age, would be "the equivalent of maybe biopsying 25-year-olds" for Alzheimer's. Hence, testing younger cattle would not necessarily reveal anything, Detwiler pointed out. Other countries have fallen into the trap "where they're testing very young animals and that's how they're building up their numbers. You wouldn't find the disease in those numbers," she stated. "Testing doesn't buy you protection."

Bovine Barricades

Protection stems from regulations, and, Detwiler observed, the U.S. was lucky because it already had import restrictions on particular countries in the 1980s for other livestock diseases; many of these countries later turned out to have BSE. "I hate to say it in the same breath, but foot-and-mouth restrictions actually prevented a lot of [BSE-risky material] from unknowingly coming to us in the 1980s," Detwiler explained of the viral disease that doesn't pose a human health threat, but which is often confused with mad cow disease. The first BSE-specific regulations from the USDA began in 1989, and they restricted imports of ruminants from BSE countries. In 1991, the U.S. banned the importation of meat-and-bone meal (as well as meat for human consumption) from those countries. In 1997, the U.S. extended the ban to include countries considered at risk for BSE: namely, the entire European continent. "That was a hard sell," Detwiler remembered of her TSE working group's recommendation to the USDA. "It was on mere risk alone," based on "where the feed moved." When BSE began showing up throughout Europe in 2001, it proved to be prudent.

The Food and Drug Administration (FDA), which monitors the safety of food products and animal feed, also erected barriers. Most notably, in August 1997 it banned most mammalian protein from ruminant feed—exceptions were given to protein from pigs, horses, milk, blood, gelatin, and leftover human food (plate waste). Completing the BSE firewall are the U.S. Customs Service, which screens goods entering the country, and the USDA's Food Safety Inspection Service (FSIS), which monitors the safety of meat, poultry, and some egg products.

These measures have made the risk of a mad cow outbreak low, concludes a November 26, 2001, report by researchers from the Harvard Center for Risk Analysis and Tuskegee University's Center for Computational Epidemiology.[8] In 1998, the USDA asked these researchers, led by George M. Gray, to assess the effectiveness of the regulations in preventing the spread of BSE. The researchers looked at the European situation, studied current rules, and visited farms, slaughterhouses, and processing plants in the U.S. They constructed a mathematical model—essentially a statistically weighted flow chart starting with infectivity sources and advancing to cattle exposure, then to slaughterhouses, and finally to feed. As a variable, they also included noncompliance with the regulations—in January 2001, the FDA found that 16 percent of American renderers and 20 percent of feed mills that are licensed (because they handle animal medicines as well) did not properly label their feed. Packaging on mammalian protein must indicate that the feed cannot go to ruminants. Unlicensed mills fared worse—41 percent did not comply. More than a quarter of renderers and unlicensed mills did not have systems in place to prevent commingling of feed for different varieties of livestock.

The team ran scenarios in which various numbers of infected cattle were surreptitiously brought ashore. "Of course it's illegal to import cows into the United States from places that currently have BSE. But we wanted to see what might happen. So we ran scenarios with the importation of one cow, five cows, ten cows, all the way up to 500 cows infected with BSE," Gray explained at a press conference presenting the study. "We ran dozens of scenarios and thousands of variations of each of those with our model, and we couldn't come up with a single situation where BSE could take hold or spread in any significant way. In every case, the disease dies out, usually in about 20 years."[9]

The Harvard assessment also factored in the 173 "missing" cows imported from the U.K. in the 1980s, which could have wound up in human or cattle food. They assumed that the cows were affected by

BSE, and using information such as the animals' age, year of import, and the date of the animal's last known sighting estimated the theoretical amount of infectivity that could have been introduced into U.S. herds.

> Our analysis concludes that there is more than an 80 percent chance that the import of these animals resulted in no exposure of U.S. cattle to BSE infectivity. Even if U.S. animals were exposed to BSE, there is a significant chance that the exposure resulted in no new cases of disease.[10]

If there were new cases of BSE, the Harvard assessment said, they would have escaped detection by then-existing surveillance, but control measures since 1997 would have halted the spread of the disease. Any new cases would not lead to an epidemic "largely because of the feed ban by the Food and Drug Administration, [which] although not perfect, breaks the loop" and would allow infected parts of a cow to enter a healthy animal's diet, Gray explained.[11] In all likelihood, the 173 unaccounted cattle from the U.K. were not infected, the USDA has concluded. They were beef animals, not dairy cows that BSE preferentially struck. Furthermore, they came from farms that had not seen cases of mad cow disease the year the cattle were born. Moreover, until the U.S. government lost track of them, the imports remained healthy beyond the average incubation period of five years.

To lower the risk of BSE further, the U.S. may eventually adopt some of the control measures used in Europe. The FDA may ban high-risk material from all rendered products. The USDA has proposed banning the use of cattle stunning devices that inject air into the skull to scramble the brain. Based on a recommendation from the FDA's Transmissible Spongiform Encephalopathy Advisory Committee (TSEAC) —a collection of TSE experts that meets two or three times a year—the government may ban central nervous system tissue from food. This change would better ensure that brain and spinal column material stay out of hot dogs, sausages, and other meat extracted mechanically from the carcass through a process called the advanced meat recovery system. This extraction method sometimes pulls out neural and spinal parts: In one test, the USDA found that 12 out of 63 beef samples contained tissue from the central nervous system.

Breaks in the Firewall

In principle, the regulatory dam erected by the government should keep BSE out of the country. In reality, however, there are several spots where leakage is possible, and some of the regulations were expressed in ways that could provide a false sense of security. For instance, before 2002 the USDA surveillance program was testing downer cows but failed to state that such cows don't include those that die on the farms—a population just as likely to harbor BSE as those that can't walk through the slaughter chutes. Dead-on-the-farm cattle tend to be old animals and often die of unknown causes. The animals could be buried on the farms (which can be difficult to do in the hard ground of winter), dumped in a landfill, or sent to renderers to become animal feed. Even if such feed were properly labeled as not fit for cows, it could still infect ruminants. The reason is that chickens may eat the feed—and "chicken litter" may be fed to cows. Chicken litter is the stuff swept off the floors of chicken houses—and that includes feed, feathers, and feces. The practice is believed to be rare, restricted mostly to on-farm use, but it is legal nonetheless.[12] As such, the Harvard assessment identified the dead-on-the-farm cattle as potential sources of infectivity, so now the USDA does try to test those animals.

Several other areas could use additional strengthening, concluded a January 2002 report by the General Accounting Office, authored primarily by Lawrence J. Dyckman. Beef products from at-risk countries aren't supposed to get into the U.S., but not all points of entry are sealed. The GAO report identified international bulk mail as one source. Thanks to x-ray technology, government officials can determine (with varying degrees of accuracy) whether a package has organic material in it. USDA inspectors once seized corned beef from Ireland, a banned country, from a container labeled "cutlery." But many more packages go uninspected. Between May and October 2001, some 1.5 million packages went through a New Jersey international bulk mail facility (one of 14 in the nation), but the USDA could only inspect about 7 percent of them. Of those they did look at, 570 (0.5 percent) contained at-risk beef products. The examination rate is low because only one or two inspectors are on duty at any one time, and each only has seconds to examine a package moving on a conveyor belt.[13]

The U.S. Customs Service has also seen discrepancies in the information that importers provide, thus permitting banned beef to enter. One importer said that beef was coming from Canada when it was actually from Switzerland. In another shipment, animal feed was incorrectly labeled as pet food. In 1999 Customs found that 21 percent of fresh and frozen beef and 24 percent of animal feed were inaccurately described.[14]

Sometimes the rules themselves permit overly broad terms to be used, preventing an accurate enumeration of the amount of banned material that might be making it ashore. After *The Wall Street Journal* printed a front-page article in November 2001 about banned feed getting into the country, Detwiler said the USDA ran a check to find out what exactly the feed was.[15] "Ever go into a pet store?" she asked. In the "dog" section, there will likely be barrels full of pig ears. "Oh man, we import a lot of pig ears," Detwiler remarked, "and they're classified as animal feed," even though they are not the kind of thing for livestock. Animal products destined for research projects, such as flavor testing, are also classified as animal feed, Detwiler said. On the other end, some meat products imported from BSE countries legally arrive as "non-species specific." Without genetic testing, it's impossible to ensure that such goods are not derived from cattle.

Michael Hansen of Consumers Union illustrated a disturbing loophole in the way the regulations are written with a hypothetical example. Suppose a BSE country wanted to export cow brains to the U.S., he said. The USDA would stop them. But the country could appeal to the World Trade Organization and re-label the brains to say, "do not feed to ruminants," in which case the material would be allowed in.

Realistically, it's not possible to prevent all at-risk material from entering the U.S.—in the future, more of the stuff will undoubtedly creep in as the amount of global trade rises and inspection capacity fails to increase with it.[16] That's why a feed ban is so critical—if any BSE prions entered the food chain, the ban would keep the malformed prion proteins from replicating to epidemic proportions. Yet the feed ban is the weakest link, the GAO report concluded, and is more permissive than those of other countries.

Most of the problem is rooted in the enforcement of the rules, especially during the late 1990s. Unlike the U.K., the U.S. did not actually test feed to see if it was contaminated with risky material, although in 2002 the FDA promised to conduct 600 such tests and will increase the number if it finds evidence of contamination.[17]

When finding a firm that is not complying with the feed-labeling rules, the FDA resorts to its only real weapon: the warning letter. The GAO found, however, that the FDA was often slow in admonishing firms: In one instance, 21 months passed between the time the FDA inspected the facility and the time it sent the letter. Many firms voluntarily comply, but checking them is problematic—in some cases, the second inspection occurred more than two years later. Even if the firms are re-inspected, the FDA has no strategy to force compliance. The FDA instead prefers to educate firms and work in cooperation with them—a strategy that seems to work, at least some of the time. In January 2001, the FDA found that about one quarter of renderers and 20 percent of licensed feed mill operators weren't properly labeling their feed or didn't have a system to keep ruminant feed separate from other feed. In April 2002, the FDA found that licensed firms that were out of compliance had fallen to at most 7 to 8 percent; unlicensed feed mills ran about twice that rate.[18]

The GAO, however, discovered enough flaws in the FDA's database—missing entries, incomplete identifiers, contradictory responses on the same form—that it recommended not using the database as a means to assess compliance. In fact, the GAO concluded that noncompliance may be higher than reported by the FDA because the FDA treated blank entries on compliance questions as if the firms followed the rules, even when other records and inspector notes suggested that the firms did not.

The FDA has promised to address the loopholes. In October 2001, it began cleaning up the database. But other fixes may be long in coming. "The FDA has been saying they're going to tighten up the feed ban," Hansen remarked, but noted that the agency, like most of the government, is bogged down by inertia and is woefully slow in acting—an opinion based on "my years of trying to get action on BSE." Hansen referred to the BSE firewall as a picket fence; because of the disease's long incubation time, we will have to wait until the latter half of this decade to know for sure that BSE didn't slip through the gaps.

Despite gaps in the firewall, the risk of BSE appearing in the U.S. is probably low. Given that sporadic CJD strikes one in a million people apparently at random, is it reasonable to assume the same goes for cattle, resulting in an American mad cow? "It's not reasonable to *assume*," Paul Brown explained. "It's reasonable to ask the question, whether spontaneous disease occurs in other mammals at the same rate. If it's the same rate, I'm not sure we'll ever find the answer. Because in order to get a statistically secure negative, I think, the numbers have to amount to several hundred thousand" tests, he said. "Nobody wants to spend the money to do 400,000 tests just to prove that spontaneous disease does or does not occur." The tests that look for prion protein are not in themselves expensive, costing less than $60 per animal, Detwiler explained. The real expense comes from storing the carcasses, collecting and shipping the samples, labor, and travel, all of which are much more of an issue in the vast U.S. than for smaller European nations.

Several decades ago, however, before the advent of immunoassays and surveillance and feed bans, evidence appeared to suggest that U.S. cows did indeed harbor some sort of prion disease. If true, then the disease produced entirely different symptoms from BSE seen in the U.K.—a distinctly American strain, but just as deadly to those who consumed it. At least, that was what some mink were indicating.

American Madness?

For the mink industry by the mid-twentieth century, fur trapping had largely given way to fur farms, which account for about 90 percent of the mink fur sold. In the wild, mink are predators, hunting for frogs and minnows in water, and rabbits and snakes on land. In captivity, most mink consume commercially prepared feed from makers such as Kellogg's and Purina.[19] In the 1940s and 1950s, however, mink ranchers formulated their own rations, making trips to fish processing plants and slaughterhouses. They blended their own cereal mixtures and often incorporated meat-and-bone meal. Mink kits begin eating feed at four to five weeks of age, while still nursing for the next week or two.

Disaster struck one mink farm in Wisconsin in 1947, when all of the adult animals began showing signs of a progressive neurological disease—including aggressiveness, incoordination, and self-mutilation—that ultimately proved fatal. Another 125 animals shipped from the

farm to Minnesota began exhibiting the same symptoms. The next outbreak occurred in 1961, striking five ranches in Wisconsin and killing 10
to 30 percent of the adult mink. Two years later, a third set of outbreaks
occurred on ranches in Canada, Idaho, and Wisconsin. Veterinarians G.
R. Hartsough and Dieter Burger of the University of Wisconsin–
Madison concluded that the disease was infectious and moreover found
a common denominator in each set of outbreaks: the feed. They examined neural tissue from the 1963 mink and found that microscopic
spongy holes peppered their brains. The scientists also noted in their
1965 report that disease was transmissible via intracerebral inoculation,
just as scrapie was, and termed it transmissible mink encephalopathy, or
TME.

Richard Marsh, the son of an Oregon mink rancher, began following
up the studies of TME after moving to the University of Wisconsin–
Madison in 1963 to work on a second doctorate. He stayed there, and
over the next four decades, he became a giant in the field and the
world's expert on this obscure prion disease. With help from William
Hadlow, the NIH Rocky Mountain Labs researcher who first noticed
the connection between kuru and scrapie, he described how the
mink—never a friendly animal to begin with—met such an aggressive,
disturbing death: "The mink vigorously attacks, almost as though fren-
zied, an object moved along the sides of the cage. Its responses to touch
and sound are exaggerated. Loud noises easily startle it. Early on, the
mink becomes careless in defecating; it deposits its feces randomly
instead of at a single site as normal mink do." Soon, it loses control of its
hindquarters and has to drag itself around on its belly. Its hyperex-
citability gives way to a stuporous expression unless aroused. In the
advanced stages, if you extend a stick, it will clamp down on it and hold
so tenaciously you could lift the animal. It will also compulsively bite
itself to the point of mutilation and may even fatally amputate its own
tail. Once symptoms start, death usually ensues in about six to eight
weeks, although in some cases the clinical course lasts only a week. "In
the end," Marsh and Hadlow explained, the affected mink "becomes
stuporous and is often found dead with its teeth firmly clamped onto
the wire mesh of the cage."[20]

Considering that scrapie existed in the U.S., contaminated sheep
seemed to be a likely source for the infection. Marsh tried experimen-

tally infecting mink with scrapie brain tissue from different sources. But none of the inoculants produced the 7- to 12-month incubation period seen when mink were infected by TME brain tissue. Most of the American strains of scrapie brains didn't produce symptoms in the mink until about 12 months after inoculation; scrapie strains from the U.K. only made one out of 65 mink sick, and that didn't happen until 22 months after the animal received the inoculation.

Then, in April 1985, when mad cows were first being discovered in England, minks on a ranch in Stetsonville, Wisconsin, started coming down with TME. The last recorded outbreak in the U.S., the epidemic wiped out 60 percent of the 7,300 mink in five months.[21] "The Stetsonville incident is especially interesting because this rancher was a 'dead stock' feeder who used mostly dairy cows which he collected daily within a 50-mile radius of his mink ranch," Marsh and Hadlow wrote.[22] The owner was one of the few remaining ranchers who created his own feed from scratch, rather than buy commercially prepared mink chow. He kept meticulous records and did not feed sheep to his mink. On visiting the farm, Marsh discovered that the rancher fed his mink several "rabies negative" downer cows—cows that acted as if they were infected with the brain-destroying rabies pathogen but did not test positive for the viral infection. Marsh wondered if these neurologically ill cows were instead harboring a transmissible spongiform encephalopathy.

By the time Marsh learned of the Stetsonville mink, the cattle carcasses were long gone, so he couldn't test the bovine for a TSE. So he tried the next best thing. He intracerebrally inoculated two Holstein steers with brain tissue from mink that died of TME. The cattle went down 18 and 19 months later. Marsh then conducted a "back passage" experiment, taking brain tissue from the dead bovine and giving it to healthy mink. Mink that got bovine brain injected into their skulls died four months later; those that were fed cattle brain lasted longer but eventually succumbed as well, at around seven months. Besides indicating that there was virtually no species barrier between cattle and mink when it came to this particular strain of TSE, the results of the experiments were consistent with the hypothesis that the Stetsonville mink contracted the disease from infected cows. Marsh and Hadlow concluded: "If TME results from feeding infected cattle tissues to mink, there must be an unrecognized BSE-like infection in American cattle and in other countries where TME has been reported." (Isolated TME outbreaks also occurred in Finland, Russia, and Germany.)

Moreover, in his experiment Marsh found that his TME-inoculated steers did not display the aggressive and staggering movements typical of BSE symptoms. Rather, just before dying, the American cattle simply collapsed in their holding pens and wouldn't get up. Whatever prion disease these cattle had, it was different from the BSE that appeared in Britain. Recognizing that an American strain of BSE could be amplified if rendered cattle were turned into cattle feed, Marsh lobbied hard to have the beef industry end the practice. In 1990, he wrote a paper for *Hoard's Dairyman*, a national dairy farm journal, that called for such action. Appearing when mad cow furor in the U.K. was reaching its peak, it created a local storm. Marsh became the source of much antipathy from the $3-billion-a-year rendering industry, which processes some 25 million tons of animal material each year. Rendering officials pleaded for Marsh to change his stance—the connection between cows and TME wasn't a slam dunk, because the rancher also fed organs from other animals rejected by feed companies. Marsh stood firm, even after the officials went to the dean of Marsh's college at the University of Wisconsin–Madison.[23]

Marsh died in his Middleton, Wisconsin, home from cancer on March 23, 1997, just before the FDA did what he had been clamoring for: It banned mammalian protein from ruminant feed.

In Case of Emergency . . .

That cows might exhibit different strains of a prion disease isn't surprising and in fact may be expected—after all, humans have five main ones (Creutzfeldt-Jakob disease, variant CJD, kuru, fatal familial insomnia, and Gerstmann-Sträussler-Scheinker syndrome); sheep have at least nine, possibly two dozen. Was Marsh correct in his theory of additional BSE strains? "Because of Dick Marsh, we were the first country to start testing downer cattle. We actually started testing downer cattle at the end of 1993 into 1994, because of that theory. We were looking for a different clinical presentation," Linda Detwiler explained. "We also started using immunohistochemistry at that time," in case this strain didn't produce the spongy holes typical of BSE in the U.K. (Immunohistochemistry relies on antibodies to bind to prion protein; color agents then bind to those antibodies, thereby visibly staining the PrP.) With all the brains examined so far, Detwiler remarked, we "still have no evidence of a BSE-like disease or another type of TSE in cattle." Back in 1994, though, the U.S. only examined 199 brains from downers—a mere 0.1 percent of the downer cattle population. The testing of 19,900 downers in 2002, all negative for prion disease, is more reassuring.

Considering that a case of BSE would scare off many Americans from beef—each person consumes some 64 pounds each year—and produce a short-term economic calamity, one would think that ranchers might be tempted to hide a mad cow. But Detwiler thought the scenario was unlikely. The system is built so that individual reporting of suspect cases is not needed: Cows for testing are taken from the various places that collect downers and cattle that die on the farms. "If we go to the plants that do the slaughter of the nonambulatories, if we go to the renderers that collect the dairy deads off the farm, if we go to the rabies labs, the veterinary diagnostic labs, anything that would have cows with neurologic signs at all, it automatically filters into this system," Detwiler explained. "Like spokes coming into a hub."

If surveillance does turn up a suspect case, Detwiler and her colleagues swing into action according to the guidelines set forth in the BSE Red Book, titled *BSE Emergency Disease Guidelines.* If an inspector from the USDA's Food Safety Inspection Service (FSIS) spots a sick cow at the slaughterhouse, he or she pulls the animal and refers it to the Animal and Plant Health Inspection Service (APHIS). Investigators examine the brain of the animal at APHIS's National Veterinary Services Laboratories (NVSL) in Ames, Iowa. They look for the telltale microscopic holes and run immunoassays to search for the presence of protease-resistant prion protein. Any suspect tissue is tested again. All this takes place within 10 to 18 days after the lab first receives the sample.

On a presumptive diagnosis of BSE, members of the NSVL hand-carry a sample and fly to the U.K.'s Central Veterinary Laboratory, widely regarded as the world's reference laboratory for BSE. The CVL can confirm the diagnosis within 24 to 96 hours. Concurrently, APHIS officers start quarantining the herd and trace the offspring and herd mates within three days. If necessary, FSIS field personnel obtain all information about the carcass—especially the whereabouts of the brain and spinal cord—trace all food items that may have come from the cow, and follow the feed trail back to the renderer.

While the U.K. lab tests the sample, the BSE Response Team would assemble. Consisting of members of APHIS, FSIS, and other officials, the team would meet in the "Situation Room" at APHIS headquarters in Riverdale, Maryland.[24] The team would then collect information from the field offices, coordinate teleconferences with various federal agencies, notify foreign embassies, and set up a toll-free number for industry representatives, the media, and the public—all by the day the case is confirmed by the U.K.[25] Just as is done in Europe, APHIS would depopulate the herd to see how much infectivity is actually there.

Pigs and Sheep

Strict enforcement of the feed ban may actually mean that BSE will never be found in the U.S., although no one can guarantee zero risk. The feed can still go to pigs and chickens, both of which make the prion protein naturally. So far, no prion disease has been found in chickens, but inoculation experiments have shown that pigs can contract it, if rather inefficiently. The good news is that pigs fed brains from BSE cattle have not contracted any prion disease. Moreover, brain tissue from those pigs did not transmit any prion disease when injected into mice, Detwiler said—an important result, because experiments have shown that animals can be silent carriers of prion disease. In November 2001, veterinarian Richard E. Race and others at the NIH Rocky Mountain Labs discovered that mice experimentally infected with hamster scrapie did not develop any clinical signs, yet their brains and spleens were still infective, capable of passing on the prion disease to other mice and hamsters.[26]

Consumer Union's Michael Hansen believes that the protocol of that U.K. pig-feeding experiment was flawed, because pigs ate only a total of 1.2 kilograms each. Although that is about the amount a pig might typically eat during its life, the dose was given over a matter of weeks. The broader question of whether pigs can be orally infected regardless of dose is not addressed. "A negative finding would be hard to interpret and would *not* mean that BSE is not orally active in pigs," Hansen wrote in a May 1997 letter to the FSIS.[27]

Hansen also pointed out that pigs might have once contracted a prion disease in an agricultural setting. In 1979, researchers examined 106 pigs at the Tobin Packing plant in Albany, New York, to determine the cause of an outbreak of a neurological disease. The brain of one of the pigs showed diffuse astrocytosis and other lesions but no obvious spongiform change, according to William Hadlow, who reviewed the slides. But because the slides were poorly stained, Hadlow could make no definitive conclusion—many other diseases induce astrocytosis. Moreover, Detwiler said, there was no evidence of the protease-resistant prion protein in the preserved samples.

Still, it's theoretically possible that BSE prions could adapt to infect pigs. In the Rocky Mountain Lab work on mice carrying scrapie without showing symptoms, the prions could change over time. A group of mice got sick two years after being inoculated with the brains of hamsters originally infected by scrapie-infected asymptomatic mice. "The scrapie [agent] seemed to have learned how to deal with this new species, and it worked much better. It replicated faster in additional rounds of mice and even became more lethal to them," Race explained. Moreover, the pathology was different for individual mice, suggesting that the scrapie prions had formed multiple strains. "Because we have further confirmed that prion disease can adapt to new species, and because we've shown that process is slow and difficult to detect, it may be time to rethink this practice" of feeding rendered cows to pigs, Race stated.[28]

Rather than investigating the theoretical possibility of a pig prion disease, the USDA is more concerned with existing ones in livestock. The government is trying to eliminate scrapie entirely—although for economic reasons, not as a matter of food safety. Many countries will only take sheep from scrapie-free countries, leaving U.S. flock owners and 11.5 million sheep shut out of much of the international market. The USDA estimates that scrapie costs the U.S. sheep industry $20 million per year in direct costs and many more millions in lost potential markets.[29]

Since the first cases of scrapie appeared in the U.S. in 1947, the government has tried several eradication programs, none of which succeeded in wiping out the disease among the nation's flocks. The aggressiveness of some plans, such as total flock destruction, encouraged some farmers to hide their scrapie cases. So in 1992, the U.S. instituted a voluntary program in which farmers can have their flocks certified as scrapie-free if the flocks meet certain health and maintenance conditions. Federal action is still taken against potentially infected herds, such as the depopulation of two flocks in Vermont in 2001. To prevent possible future contamination from the Vermont farms, all the material that could burn—wood, harnesses, compost, manure piles—were collected and incinerated. Solid surfaces were disinfected with commercial-grade chlorine bleach. In lambing areas, 6 inches of soil was removed; for the rest of the pastures, only limited access will be allowed for five years. With newer diagnostic tests that can be performed on live animals (analyzing a bit of the lymphoid patch on the sheep's third eyelid or possibly even sampling their urine), Detwiler thinks it is conceivable that the U.S. could be a scrapie-free nation someday. Given past eradication efforts around the world, it would not be surprising if the effort failed.

But there is one prion disease of animals running completely rampant in some areas of the country—some places are so hopelessly contaminated that vets don't even bother taking precautions to keep prions from potentially getting into the soil. It's a uniquely North American affliction, and deer hunters aren't happy about it.

CHAPTER 11

Scourge of the Cervids

Chronic wasting disease of deer and elk, once confined to a patch in the Rockies, spreads across the nation.

The eradication zone, covering some 411 square miles, lies about 40 miles west of Madison, the capital of Wisconsin. There, in the southwest region of the state, thousands of white-tailed deer live—or rather, used to live. Starting in June 2002, the Wisconsin Department of Natural Resources instituted special hunting periods in hopes of wiping out an estimated 25,000 deer. All dead deer were taken to one of two registration areas, where state employees in protective suits and gloves dragged carcass after carcass from pick-up trucks and lifted them onto plastic-covered picnic tables. A tooth was pulled from each animal to determine its age. Then the gutting began—first a butcher's knife to slice through the fur, then a hacksaw to sever the head, which was double-bagged and sent for testing. Instead of being enjoyed as venison, the body was incinerated or, possibly, thrown into a giant vat of boiling lye.

There was no guarantee that the special extended hunting season would get all the "mad deer." By January 2003, only about 11,000 deer had been bagged. Thomas Givnish, a botanist and an expert on the ecology of diseases at the University of Wisconsin–Madison, doubted that such an uncoordinated hunt would truly eradicate the population in the hot zone. More aggressive attempts, such as using dogs to funnel

Carcasses of deer from Wisconsin's eradication zone are loaded for disposal by Mark Schmidt, left, and Steve Krueger, both with the state's Department of Natural Resources, in March 2001. The deer were decapitated so that their brains could be tested for chronic wasting disease. (*Andy Morris/AP Wide World Photos.*)

the whitetails to areas where professional sharpshooters in trees can pick them off, should have been implemented, he suggested. And the eradication zone should have been extended by 100 square miles "to take into account the scale of the natural movement of deer," whose ranges can extend 10 miles, Givnish said.[1]

Although Wisconsin only found 50 sick deer by the end of 2002, state wildlife officials and scientists remain deeply concerned. The incidence levels ran near 2 percent. But unlike TSEs in sheep and cattle, the deer are wild, presenting the potential for an uncontrolled spread to the state's 1.6 million other white-tailed deer—and possibly to the rest of North America's. And their prion disease spreads from one deer to another even more aggressively than scrapie does among sheep.[2] What's astonishing, too, is the fact that chronic wasting disease, or CWD, has reached Wisconsin in the first place. It managed to move east across the Mississippi River from its presumed starting point near Fort Collins, Colorado.

There, in 1967 at the state's Foothills Wildlife Research Facility, CWD made is first recorded appearance, in captive mule deer that were being maintained for nutritional studies. As the name of the disease suggests, affected deer lose weight over the course of weeks or months. They often become excessively thirsty, driving them to drink large amounts of water and consequently to urinate a great deal; they also start slobbering and drooling. In some cases, the esophagus loses tone and becomes flaccid. (That probably leads to aspiration pneumonia, a common condition in the terminal stages of the disease.) They stop socializing with fellow deer, become listless, and have blank facial expressions. From the start of clinical signs, death ensues in about three to four months, although some expire within days and others in about a year.[3]

The Fort Collins facility became a CWD death trap. Between 1970 and 1981, 90 percent of the deer that stayed more than two years died from the disease or had to be euthanized after the onset of symptoms. In 1980, the scourge appeared just outside Colorado's border, at the Sybille Research Unit in southeastern Wyoming, 120 miles northwest of Fort Collins. The two facilities had shared deer, thus indicating that the disease was infectious—even to a different species: Soon, the elk at the facilities contracted the disease. (Deer and elk belong to the same family and are called cervids.)

For years, researchers thought the disease resulted from nutritional deficiencies, poisoning, or stress from confinement. The unknown cause was a challenge to Elizabeth S. Williams. Interested in pathology and wildlife diseases, she had come to Colorado State University for her Ph.D. after receiving her veterinary degree. One night in 1977, she looked at brain slices from CWD animals and saw that the tissue was full of microscopic holes. "I happened to be taking a course in neuropathology and studied a lot of brain lesions," she recalled.[4] The holes were unmistakably scrapie-like. "Not many things cause that," Williams said. With colleague Stuart Young, she published a paper in 1980, pointing out that CWD was related to spongiform encephalopathies.[5] (One speculation is that CWD originated from scrapie: Sheep with signs of scrapie were reportedly seen near the cervids. More convincing are an immunoassay study reported in late 2002 that found no distinction between the prions from scrapie and CWD tissue, and an inoculation study described in 2003 that found no neuropathological differences between elk infected with CWD and those experimentally infected with scrapie.)

But unlike BSE in cows or vCJD in humans, the cervids weren't getting sick from their food. The epidemiology suggested that CWD behaved more like scrapie in that it spreads horizontally—although how, no one really knows. The prions could lurk in the cervid's urine. During rutting season, deer bucks may lap up the urine of dozens of does to find out which are in heat. In elk, females lick males that have sprayed themselves by aiming their urine forward, soaking their bellies and even their necks when their heads are lowered. Saliva could be a vector, too; in both deer and elk, individuals meet and greet by licking each other's mouths and noses, thus exchanging drool. Ranched elk may also swap saliva when they feed in close quarters, drooling as animals do at mealtime. It's also possible that sick animals shed prions on the ground via their feces, urine, and saliva, thus giving other grazing animals an opportunity to take in the pathogen. By 1985, veterinarians had discovered CWD in free-ranging deer and elk, generally within about 30 miles of the two facilities. Whether the disease originated in the wild and spread to the captives, or vice versa, is not known. But it is clear that the two populations had plenty of time to interact. Especially during mating season, wild cervids nosed up to captives through the chain-link fence.

Both the Sybille and Fort Collins pens tried to eradicate CWD. At Sybille Canyon, all the deer and elk were killed in the main area, which was put off limits to cervids for one year. (The animals in the outlying areas, where CWD did not appear, were left alive.) Deer and elk were then reintroduced; four years later, they started coming down with CWD. The Fort Collins facility made more aggressive attempts to clear CWD. Researchers first killed off all the resident deer and elk. Then the soil was turned, and structures and pastures were sprayed repeatedly with swimming-pool chlorine. The whole area was kept free of cervids for a year. Then 12 elk calves were brought in. A few years later, two of those elk contracted CWD. "The extensive disinfection procedures followed at the Fort Collins facility should have been adequate to greatly reduce exposure of cervids to the agent," Williams and Young concluded.[6] Perhaps cervids are extremely sensitive to CWD prions, or perhaps the calves were actually already incubating the disease when they were collected from the wild. In any case, both regions today are hopelessly tainted that taking precautions against further contamination is pointless. One vet admitted he simply hosed down his pick-up truck after delivering CWD animals to Williams's lab, now at the University of Wyoming, in Laramie.[7]

For nearly four decades, CWD remained an obscure disease confined to northeastern Colorado and southwestern Wyoming, in an area of about 15,000 square miles. The research facilities have stopped trading captive animals. ("They're only allowed out to come to my necropsy room," Williams quipped.) The 14,000-foot-high mountains and other natural barriers have kept the wild deer and elk from spreading CWD easily, although in 2001 a wild CWD deer turned up in a neighboring county in southwestern Nebraska, thereby extending the endemic range. The incidence of CWD among the cervids averages about 4 to 5 percent but has reached 18 percent in some areas. Government sharpshooters have culled thousands of deer and elk to thin the herds and slow the spread.

There was, however, a quick means to transmit CWD out of the endemic area: along the roads, in a truck.

Out and About

Some 2300 ranches holding 160,000 elk dot the U.S. and Canada. Elk don't need a lot of room or food, so small ranchers turned to them as a source of extra income. Besides the meat, they could also sell the antlers, which is marketed as a supplement in vitamin stores ("velvet antler") and as an aphrodisiac in Asia ("velvet Viagra"). Stocking various ranches meant trading the animals, some of which evidently looked healthy but were incubating CWD. (It takes about 20 to 30 months on average for symptoms to show.) The first farmed elk to display signs of CWD occurred in 1996 on a ranch in Saskatchewan, Canada. A year later, a South Dakota ranch found a CWD elk in its herd. By 2001, some 20 ranches reported cases across six states (Colorado, Kansas, Montana, Nebraska, Oklahoma, and South Dakota) and two Canadian provinces (Alberta and Saskatchewan). Elk breeders and state and federal officials took aggressive measures so that most of the infected herds have been depopulated.

Still, it may have been too late—the transport of incubating cervids has evidently spread CWD to wild populations in those states as well as in New Mexico and Minnesota. The actual numbers of deer and elk with CWD is only slowly becoming clear. In a 2002 House of Representatives testimony, Michael W. Miller, a CWD expert with the Colorado Division of Wildlife, stated that "there appear to be two relatively distinct CWD epidemics occurring in North American cervid populations." One was the endemic region of wild cervids in the contiguous area bounded by Colorado, Wyoming, and Nebraska. "The other epidemic is occurring in a relatively small number of farmed elk herds scattered across the U.S. and Canada, with apparent spill-over to local populations of free-ranging deer."[8]

The unwitting transport of sick animals may also explain how CWD spread across the Mississippi River to Wisconsin and Illinois. Just how the white-tailed deer, the most common type in the eastern U.S., contracted it is unknown. "I don't think we can answer how wild deer in Wisconsin got it," Elizabeth Williams remarked. Possibly, the free-ranging deer contracted it from the farmed elk or deer. Or perhaps some captive deer escaped. "Whitetails can jump and can weasel their way out" of pens, Williams explained, noting that states are discussing doubling up the fencing or electrifying them. The disease could also stem from the cervids' contact with scrapie sheep. Alternatively, CWD may arise spontaneously every so often, just as sporadic CJD does, and become infectious, as kuru did. Continued epidemiological studies of the area may pinpoint a source, although if it turns out that CWD had been lurking in the area for the last decade, it may be impossible to ever know how it got started. "By the time these problems are discovered," Miller commented, "they have probably been sitting there for decades. That makes it difficult to go back and retrace how things came about."[9]

The number of infected cervids isn't completely known. "Animals showing mid-stage clinical disease represent the 'tip of the iceberg' with respect to the overall rate of infection in the population of interest," Miller stated in his congressional remarks. In 1996, researchers began using immunohistochemistry on a part of the brainstem called the obex and on the tonsil tissue to determine whether a cervid is infected. "We know even these [immunohistochemistry]-based estimates of CWD prevalence are still a little low," he testified.[10]

Whether the massive killing project in Wisconsin can wipe out CWD is "a really difficult call," Williams explained, because the CWD prions may have irreparably contaminated the environment, just as they did in Colorado and Wyoming. (Wisconsin officials originally planned to landfill the carcasses—which would have been a huge mistake. Given the estimated incidence of CWD, a massive prion protein load would have been put into the ground, perhaps permanently contaminating the area, or worse, eventually leaching into the groundwater.) "The idea is to find a fairly small focus and get rid of all the animals in the area. That might stop CWD from getting established," Williams stated. A rapid spread is possible in Wisconsin because the deer population in state's southwest corner is dense: Thomas Givnish noted that it runs about 50 to 100 deer per square mile, or ten times that of the endemic area in Colorado. "The alternative is to do nothing," Williams noted, and then "you know it's going to be established." The reproductively prolific nature of white-tailed deer, wildlife managers believe, should restore the population in a few years—although considering the persistence of prions, officials may only be able to keep CWD in check, rather than eliminating it.

Unfortunately, the start of 2003 brought some disconcerting news. In January, tests showed that five deer killed during the fall hunt just outside the eradication zone had CWD. Worse, one of the infected deer was found north of the Wisconsin River, which researchers had hoped might serve as a natural boundary. As a result, the state may have to widen the original eradication zone.

For outdoorsmen, CWD has proved to be a major threat to a cherished way of life. More than just venison, the hunting season means weeklong bonding among friends and relatives. Because many have been raised to kill only what they can eat, Wisconsin's request that they shoot as many as they can fills them with ambivalence. To encourage hunting in the fall 2002, the state Department of Natural Resources bought radio ads, urging hunters make Wisconsin "CWD free in 2003" and playing a jingle by a local group, Bananas at Large: "Stay out on da trail./Make a brighter future for huntin' White Tail./Bag 'em, tag 'em, drag 'em/freeze 'em, test 'em, fry 'em./I ain't afraid of no twisted little prion."[11] (Note that "prion" has to be mispronounced as "pry-on" to make the rhyme just barely passable.) Given that kills had fallen thousands short of the target of 25,000, state officials decided in February 2003 to bring in sharpshooters to take out more of the deer in the eradication zone.

Venison and Beyond

No one knows whether CWD can spread to humans. In 2000, Williams, Miller, Byron Caughey, and other collaborators reported on in vitro experiments that mixed CWD prions with normal prion proteins from cervids, humans, sheep, and cows. The CWD prions readily changed the normal cervid PrP to the pathological form—but had a hard time converting human prion protein. The process was extremely inefficient—less than 7 percent of the human prion protein was changed by the CWD prions. And the conversion rate for sheep and bovine PrP was not much different.

The downside of the results is that CWD prions convert human PrP about as efficiently as BSE prions do. And because BSE has infected humans, one might argue that CWD poses a similar risk. But because the dosage matters—beef is far more popular than venison—CWD doesn't present quite the same challenge as BSE does. Moreover, test-tube studies make poor substitutes for cells—in vitro results often differ significantly from in vivo data—so drawing any firm conclusions isn't possible.[12] About the only solid statement to be made is that a significant CWD species barrier exists between humans and cervids.

So far, no one has documented a case in which CWD has definitely spread to a human, although the Centers for Disease Control and Prevention did investigate the cases of three young venison-eaters who died of sporadic Creutzfeldt-Jakob disease after 1997. All were under 30 years of age, which is exceedingly rare for CJD victims. In fact, through May 31, 2000, only one other case of sporadic CJD occurred in a young person since surveillance in the U.S. began in 1979.

The first was a 28-year-old cashier whose mother said she ate deer and elk meat as a child, from her father's hunts in Maine. The second was a 30-year-old salesman from Salt Lake City who had been hunting regularly since 1985. The third was a 27-year-old truck driver and avid hunter from Oklahoma who harvested deer at least once a year. But none of them ate venison from the endemic area. The CDC tested the 1037 deer and elk taken during the 1999 hunting season from the regions where the victims' meat had come from; all turned up CWD-negative. Pierluigi Gambetti's National Prion Disease Pathology Surveillance Center in Cleveland examined the brains of the CJD victims and found no distinguishing features or unique prion protein signature, as might be the case for a new CJD strain.[13]

The CDC also looked at a 25-year-old prion disease victim from southeastern Wyoming who ate local venison but found he'd had the genetic mutation for the inherited Gerstmann-Sträussler-Scheinker syndrome. Two other young prion disease victims from neighboring counties, who fell ill within months of each other, were also examined, but no link with CWD could be established. One seemed to have died from GSS, whereas the other did not consume venison. States with CWD do not have a higher incidence of CJD, either.

In the summer of 2002, the Wisconsin health department asked the CDC to review the autopsies of three outdoorsmen who had died of neurological illnesses. The three friends had all participated in wild deer and elk feasts at one time or another in the late 1980s and early 1990s—one died in 1993, the other two in 1999. Only one had a CJD diagnosis; another succumbed to Pick's disease, a rare ailment but about 20 times more common than CJD. The cause of death of the third man was unlisted. In its investigation, the CDC could not tie these three cases to CWD.[14]

But just because health officials haven't been able to link these suspicious cases to chronic wasting disease doesn't mean that the connection can be ruled out. The CDC concludes that surveillance of human prion diseases, as well as strain-typing and lab analysis, is critical to determine if CWD can jump to humans. "It is generally prudent to avoid consuming food derived from any animal with evidence of a TSE," the CDC states. As such, hunters in the endemic area around Fort Collins can have their kills tested by dropping off the deer heads in strategically placed drums.

Still, CWD prions could find their way into human mouths. Given the prion's persistence, it would be difficult to remove all the CWD prions from hunting knives used to gut the kill. Washing and wiping the blade down will do little, and the knife could contaminate other meat. (Wisconsin's Department of Natural Resources recommends soaking knives in a bleach solution for an hour.) Local venison butchers could also spread CWD prions by giving customers ground meat from pooled scraps, some of which could have come from CWD-positive kills. Worries about CWD have cut into Wisconsin's $1.5-billion-a-year hunting industry; license applications had dipped 25 to 30 percent by the fall in 2002.[15]

Scientists are still trying to determine if CWD poses a threat to domestic livestock. In an ongoing experiment begun in 1997 by Amir Hamir and his colleagues at the USDA's National Animal Disease Center in Ames, Iowa, 13 Angus beef calves were intracerebrally inoculated with brain suspensions from CWD mule deer.[16] Three became infected about two years after inoculation, two more nearly five years after. The sick cows did not display the usual BSE symptoms of aggressive, uncoordinated action. Aside from some weight loss, "the symptoms were very vague, nothing like scrapie in sheep or BSE in cattle," said Hamir, noting that judging symptoms shown in an experimental setup is difficult because the animals become hard to handle since they are so bored and may become lame since they must live on a concrete floor.[17] After they were euthanized, their brains were tested and came up positive for the toxic form of PrP.[18] None of the bovine showed the characteristic lesions seen in BSE cattle, Hamir explained, and there was not much spongiform change at all. Hamir and his colleagues began repeating the experiment on 14 calves in November 2002, this time with the brains of CWD white-tailed deer. Williams herself started an oral inoculation experiment in the summer of 2001, injecting CWD brain material into the throats of calves to see if they would develop a prion disease.

Under more natural conditions, however, bovines have not contracted CWD so far. Williams has kept cows in contact with CWD-infected animals, and more than five years on, the cows are still healthy. Bovines kept with decomposing CWD carcasses or isolated in pens that once housed CWD cervids have also remained prion disease–free. That's good news for cows grazing in pastures, which commonly find themselves in the company of wild deer.

Whether CWD poses a threat to non-ruminants is unknown—there have been no transmission studies yet to see whether pigs or chickens could contract CWD if infected cervids were turned into feed. To prevent the possibility, the FDA in November 2002 decided it would prohibit renderers from using deer and elk that test positive for CWD and strongly recommended against using those that had come from an area considered to be an endemic source of infection. (Cervid carcasses represent only a small fraction of the 50 billion pounds of material turned into feed every year.)

If American livestock so far seem to be safe from CWD, the same cannot yet be said of other animals. If a CWD deer dies in the forest and nobody is there to see it, rest assured that there are plenty of coyotes, bobcats, and other carnivores that will gladly scavenge what remains of the wasted carcass. And during the clinical phase, CWD animals undoubtedly make easier targets for predators. So far, there's no evidence that members of the canine family can get a prion disease. But felines can. Transmission studies with mountain lions have begun, and on those rare occasions that local mountain lions die for unknown reasons, their bodies do find their way to Elizabeth Williams' pathology table.

The U.S. has implemented several steps in an attempt to keep prion diseases out of the food chain. Many argue that those steps are insufficient or belated, but so far they seem to be working. Food, however, is not the only way to catch a prion disease. The history of CJD has shown that we can catch it from each other, via the medical instruments we share during surgery or from organs that we donate. Considering all of those who lived in and traveled to the U.K. in the 1980s, and all those in other countries that imported infected feed, there are many out there who could be incubating a prion disease—perhaps millions of people, if not tens of millions. That's a lot of ammunition for what Paul Brown and others have termed "friendly fire" in medicine.

CHAPTER 12

Misadventures in Medicine

Prion diseases spread to humans through medical mishaps.

Absent-mindedly, I leaned against the doorframe. "I wouldn't do that," admonished Cynthia Cowdrey. As the chief neurohistopathologist in Stephen DeArmond's lab at the University of California, San Francisco, she spends a lot of her days thinly sectioning brain tissue, both human and animal. She relies on a microtome—basically a laboratory-grade deli slicer that employs a super sharp razor to cut material pushed forward by the turns of a screw. There's no splattering, but considering how small and finely sliced the sections can be—gossamer wisps of hamster cerebellum, hippocampus, and caudate nucleus have been mounted on slides and sit in an adjacent box—pieces of brain could get flicked in the air. "Sometimes I wear a mask," Cowdrey said. "I don't like the idea of little bits of tissue that might float in the air and get snuffled up my nose."[1] Or possibly getting stuck on a doorframe and then hitching a ride on a visitor's jacket.

Anyone who has scrambled eggs in an ungreased pan knows how much scrubbing is needed to clean up the mess, thanks to all that protein in the eggs. Proteins are often extremely sticky, and prion protein in tissue is no exception. Given its near invincibility to harsh chemicals and high temperatures that would wipe out other pathogens, the scrapie form of prion protein poses a sterilization challenge. In a lab where it is extracted, Shu Chen of Case Western Reserve University explained that "anything that goes in cannot come out. So you mostly wear disposable clothing."[2] Adhesive-coated floor pads lay like suburban doormats to grab any prion protein that might be tracked out by footwear. (Labs handling pathogens must conform to biosafety ratings that range from 1 [minimal] to 4 [strict]; prion labs must meet biosafety rating 2 requirements.)

The tenacity of the prion protein requires a lab staff to be cautious. Byron Caughey of the National Institutes of Health's Rocky Mountain Labs mentioned that at one meeting, participants considered providing miniature guillotines in labs so that researchers who accidentally cut themselves would have the option of lopping off the finger before the deadly agent spread.[3] Such a measure would be extreme—and stands in marked contrast to the early days of transmissible spongiform encephalopathy research, when many scientists didn't think twice about handling infected brains with their bare hands. Early ignorance about the prion's persistence unfortunately led to tragedy—namely, the development of a TSE because of iatrogenic transmission, which is the inadvertent spread of the disease when people visit the doctor.

Surgical Spread

By the early 1970s, Carleton Gajdusek, Joe Gibbs, Paul Brown, and the other TSE investigators at the National Institutes of Health in Bethesda had begun suspecting that surgical contamination with the TSE agent was possible. Standard autoclaving—heating instruments with steam under pressure—goes on for about 15 minutes at 121° C (250° F), and that's good enough to kill fungi, bacteria, and viruses. Most succumb to just a minute's worth of boiling, let alone a quarter of an hour at an even higher temperature. But the rogue prions can survive, certainly weakened but possibly still able to infect. Such an infection may have caused the death of a 54-year-old neurosurgeon. The autopsy showed that he died of Köhlmeier-Degos syndrome, a rare blood-vessel disorder that produces excessive clots; however, he also displayed signs of a neurological disorder. In 1973, Gajdusek and Gibbs examined his brain and were able to determine that he had been suffering from a

TSE, after inoculating lab animals with a sample of the surgeon's brain. They did not rule out the possibility that he had caught the disease from a patient and pointed out the hazard for those who handle brains, especially because Creutzfeldt-Jakob disease was often misdiagnosed as Alzheimer's disease.[4]

But no medical or lab personnel has ever been conclusively shown to have contracted a prion disease—even those TSE veterans like Paul Brown, who admitted that "over the years I have certainly taken infectious material orally and been inoculated with it" during his four-plus decades in the field. "I've been contaminated, no question," he said. "You've got a needle, you've got a knife, you're cutting into the brain, and accidents happen."[5] Brown's story isn't unique: Many TSE pathologists privately acknowledge having been stuck with a sharp instrument previously used to slice infected tissue. They're generally unperturbed about such accidents—after all, it's not as if someone plunged a prion-coated metal shaft into their brains. That experience, alas, befell two other people.

In November 1976, when the then 53-year-old Gajdusek was days away from picking up his Nobel prize medal in Stockholm, he received upsetting news from Dr. Christopher Bernoulli of Zurich. Two young epilepsy patients were dying from what appeared to be CJD after undergoing diagnostic neurosurgery. Epilepsy results from uncontrolled activity of neurons that produce bursts of electrical energy leading to seizures. Drugs can control the symptoms in some epileptics. If only one patch of neurons is causing the seizures, then surgically removing the offending part of the brain can effect a complete cure. To find the defective area, doctors insert several electrodes into the brain. The procedure—called stereotactic electroencephalography—may call for the metal rods to remain stuck in the head for several days. During that time, they pick up the brain's electrical signals, and using the data, a surgeon can triangulate the position of the epileptic center.

One of Bernoulli's patients was a 69-year-old woman who had first come to him in May 1974 when she began suffering from CJD. To monitor her brain waves, Bernoulli inserted six-millimeter-wide electrodes through her skull and left them there for two days in September. After the procedure, the stainless steel and silver rods were cleaned off with benzene, a 70 percent alcohol solution, and formaldehyde vapor. Later that year,

Bernoulli used two of the electrodes on a 23-year-old woman and on a teenage boy. They developed CJD 16 and 20 months after they were exposed.[6]

To prove the link between the elderly patient and the young CJD victims, Gibbs retrieved the electrodes—it was now two years after the time of contamination, and the instruments had been sterilized at least three times—and implanted them into chimpanzees. The animals subsequently contracted the spongiform encephalopathy.[7] "These two cases represent the only fully proven instances of iatrogenic CJD" via surgical instruments, Brown concluded.[8] Five other cases—one in France and four in the U.K.—are likely to have been contracted in the same way.[9] Retrospective studies suggest that in the 1950s, for instance, three patients in the U.K. died from CJD after having operations at the same neurosurgical unit and by the same neurosurgeon during an eight-month period.[10]

Clearly, surgical instruments could be a prion-harboring nightmare. Charles Weissmann of the U.K. Medical Research Council's Prion Unit at the Institute of Neurology in London and his colleagues have been exploring just how well prion protein can bind to medical equipment. "We developed a system using stainless steel wire as a model for a surgical instrument and a mouse as a recipient," Weissmann explained.[11] The researchers placed the wires in contact with scrapie-infected brain. It took only five minutes for the wires to acquire a lethal amount of infectivity. Then the researchers "operated" on healthy mice, inserting the steel wires into their brains for 30 minutes or 120 minutes, as might be done during neurosurgery. Sure enough, the mice acquired the prion disease in both cases. Gold wires and various kinds of plastic (polystyrene, polypropylene, and polyethylene) also tightly bind the protein and transmit the disease.

Weissmann also tested various sterilizing procedures to find out which were the most effective against the tiny pathogen. An hour's worth of 10 percent formaldehyde vapor proved to be insufficient—no surprise there, given the two young CJD victims. A one-hour soak in sodium hydroxide worked, but sodium hydroxide is also extremely harsh—which is why it is terrific in clearing clogged drains—and can damage the delicate surfaces of some instruments. Guanidinium thiocyanate was less damaging, but instruments had to be soaked for 16 hours before the infectivity was removed. "We have to repeat the pro-

cedure with variant CJD and presumably have to do experiments in primates," Weissmann added.

A surefire way to decontaminate prion-coated devices is to cook
them in an autoclave at an extremely high temperature, but the
instruments have to be robust enough to withstand the extreme heat.
"Seven minutes at 137° C works, but that almost certainly isn't being
used generally in hospitals," Weissmann remarked. Rather, surgical
implements are usually heated for 15 minutes at 121° C (250° F). It
would seem that CJD should be spread in hospitals, but this doesn't
seem to be happening. Although standard sterilization cannot eliminate all infectivity, experiments with hamster brains showed that
heating at 121° C for five minutes can knock down the infectivity by a
factor of 1,000,000, Brown said. "Why haven't we seen scads of
patients who have had neurosurgery in the past who have come down
with CJD? Well, the reason is that although the sterilization procedures have not been ideal, they've been adequate. That's the most logical explanation," Brown reasoned. "The fact is, we only know of five
patients in the whole freaking world where we can trace neurosurgery
as the cause."

Still, the inability to guarantee that instruments are 100 percent
TSE-agent-free leaves hospitals at risk for unwittingly transmitting
CJD. In England, at the Middlesbrough General Hospital in Teesside, a
patient had a brain biopsy in July 2002 after five neurologists were
unable to diagnose her illness. Two weeks later, a pathologist examined
the extracted tissue and determined that the woman had CJD. But
because CJD wasn't suspected until later—she did not display any of
the typical signs—the hospital did not quarantine the instruments.
Instead, they used them on 24 other patients.[12]

At the University of Pittsburgh Medical Center Presbyterian, a man
had surgery to treat a neurological condition in April 2001. He died in
early 2002, and the autopsy revealed that he had CJD. The instruments
used on that patient may have been used on some 4000 other patients
at the medical center. Both the state health department and the medical center voluntarily sent letters to those who may have been
exposed.[13] Although the estimated risk of contracting a prion disease
via surgical instruments is considered minuscule, there is no diagnostic
test for CJD, and all of the patients will have to wait at least 10 years
before they can be certain that they weren't infected.

Deadly Eyes

Donated organs, as well as surgical instruments, can transmit prion diseases; in fact, such cases have been much more common. In 1971, a 55-year-old woman went to the College of Physicians and Surgeons of Columbia University in the Washington Heights section of Manhattan. The patient was to receive a new cornea to replace a defective one that tended to cloud over in the morning and cast halos around lights. The donor was a man, also 55, who had just died of pneumonia after two months of suffering from memory lapses and muscle trembling. An autopsy revealed that he had CJD. By then, of course, surgeons had harvested one of his corneas and had given it to the woman.

For a time, the operation seemed to have been a success: the corneal graft worked perfectly, transmitting light rays cleanly back to the woman's retina. Unknown to anyone, the graft was also transmitting CJD prions. Over the next 18 months, the pathological protein worked its way around the eyeball to the optic nerve, then traversed the nerve into the occipital lobe at the back of the brain. The woman became uncoordinated, and her muscles jerked and shuddered. She began to drool, lost her ability to speak, and lapsed into a vegetative state. Twenty-six months after the transplant, she died. Two other patients, one in Germany and one in Japan, may also have contracted CJD from infected corneas.

Hazardous Hormones

In 1976, three years after the first corneal CJD recipient died, a startling thought hit Alan Dickinson, the renowned scrapie geneticist at the Neuropathogensis Unit in Edinburgh. Lying in bed, unable to fall asleep, he realized that a particular medical treatment— the administration of human growth hormone—could spread CJD widely. Generated by the pituitary gland—a bean-sized organ deep in the brain— the hormone governs growth, metabolism, and maturation. Isolated in the 1950s, the hormone, scientists realized, could be given to boost the height of children stunted by disease or genetics. Although it couldn't make giants out of dwarfs, it could add several inches to a child's height and at least bring him or her close to average. Today, the hormone is made via genetic engineering—the DNA encoding the hormone is spliced into bacteria, which then make the hormone. But early on, before the advent of recombinant DNA technology, the only source of pituitary glands was the morgue.

In 1963, the NIH created the National Pituitary Agency to collect pituitaries from cadavers and extract the residual growth hormone. Typically, 5000 to 20,000 pituitaries were processed at a time (one processor estimated it took 50 corpses to supply a child's hormone needs for a year).[14] The parts of the processed material rich in growth hormone were combined from several batches. The hormone was then shipped to pediatricians around the country for administration. Some 8000 U.S. children received the cadaveric hormone up until 1985, when the recombinant form replaced it.

Britain had instituted a similar program, and Dickinson realized that the processing methods used to concentrate growth hormone from all the pituitaries might not be able to inactivate the CJD agent. He notified the U.K. Medical Research Council of his suspicions on October 5, 1976. The council had Dickinson test his theory—he would mix a normal human pituitary gland with scrapie-infected mouse brain, extract the hormone, and then inoculate it into test animals. "When he finally did the test, his conclusion was that the processing eliminated all infectivity," Paul Brown recalled. "That was the irony of it. He was the first one to think it possible and then do the experiment." Dickinson didn't use the entire sample, and Brown noted that some infectivity could be lurking in the remaining portion. Sure enough, when Brown and others conducted a more rigorous experiment in 1991, "scrapie-infected pituitaries did in fact transmit disease to a few of several hundred inoculated hamsters."[15]

By then, no experimental proof was necessary—in 1985, Stanford University pediatric endocrinologist Raymond Hintz notified the NIH of a young CJD patient. To compensate for his under-active pituitary, the patient began daily growth hormone injections in September 1966 at age two and continued them until July 1980. By his twentieth birthday in 1984, he had grown to 5' 4". He began to suffer from dizziness that spring, and by September had progressed to slurred speech and incoordination. After his death in November 1984, an autopsy confirmed CJD. "With the speed that is almost never encountered in government reactions to potential problems," Brown wrote, "Mortimer Lipsett, Director of the NIH Institute responsible for overseeing the pituitary treatment program, held advisory meetings and within 2 weeks notified pediatric endocrinologists around the country to be on the lookout for unexplained neurologic deaths in their patient population."[16] It turned out that two previous deaths just weeks before—one of a 34-year-old Dallas music-store clerk, another of a 22-year-old Buffalo resident—were hormone-triggered encephalopathies, although the physicians had failed to make the CJD diagnosis because the victims were so young. The NIH pulled the plug on the pituitary program, and the cadaveric hormone was withdrawn from the market in the spring of 1985. (The recombinant growth hormone became available later that year.) Other nations followed suit, and reviews of past neurological deaths in young people started turning up hormone-triggered CJD.

The NIH could have known about the CJD risk of cadaver pituitaries years earlier, argued Emily Green in her May 2000 story for *The Los Angeles Times*.[17] She found a paper trail, tracing Alan Dickinson's warning to the U.K. Medical Research Council and the Council's subsequent note to the NIH. Colin Masters, a visiting Australian pathologist, responded in a letter to the MRC on May 8, 1978, that the pituitary from a CJD victim would be expected to be contaminated. Apparently, Masters didn't notify anyone in the pituitary program, thinking that the authorities should have already known about it from the medical press. Moreover, Green reported that the leading producer of the hormone, the late Alfred E. Wilhelmi of Emory University, decided not to adopt an expensive gel-filtration method developed in Sweden that might have resulted in a safer product.

Whether the CJD risk could have been anticipated and the pituitary program stopped earlier are debatable—Brown maintained that the risk was unforeseeable. Besides, Dickinson failed to prove his speculation, so there was no evidence. Once the first deaths occurred, however, Gajdusek, Gibbs, and Brown became worried that the world could be sitting on a time bomb of future CJD cases because of the growth hormone therapy. Fortunately, a widespread epidemic has not materialized. As of June 2002, 29 individuals in the U.S. have contracted CJD via the hormone—or about 0.36 percent of the recipients. (Researchers estimated that at least 140 infected pituitaries may have been processed and distributed among the lots made between 1963 and 1985.[18])

Any final tally, however, may be two decades away. "In terms of the population treated with growth hormone, it is too early to provide reassurance that they're not going to get the disease, even though time is passing since the last potential exposure," Robert Will of the U.K. CJD Surveillance Unit explained of pre-1985 hormone recipients. "The chance is getting less and less with time, but it'll be a long time before you can be sure that they're not going to get it." Most cases took about 12 years to incubate, but much longer times are possible. In May 2002, Dutch researchers reported that a 47-year-old man had died of CJD 38 years after receiving a single shot of contaminated hormone, given to determine whether his delayed growth was the result of hormone deficiency (it wasn't, so he didn't need any treatment).[19]

Differences in processing methods, donor screening criteria, and just bad luck may explain why other countries have a higher incidence rate than the U.S. The U.K. inoculated nearly 2000 people with human growth hormone retrieved from the pituitaries of 940,000 corpses between 1959 and 1985. Health officials there had seen 40 cases by June 2002, about 2 percent of recipients. (A judge ruled that the Department of Health was negligent for not heeding Dickinson's warning, and that officials should have stopped all such treatments after July 1, 1977. This ruling opened the way for compensation payments to victims' families.) France has the most, with 90 cases out of 1260 recipients, a 7.1 percent rate; there, all victims were treated with growth hormone between 1983 and 1985. Japan treated 5000 children, but none have developed CJD. Worldwide, as of June 2002, there were 161 CJD cases from growth hormone, as well as 4 cases in Australia from gonadotropin, a hormone used to promote fertility.[20]

Patch Full of Prions

Unfortunately, human growth hormone wasn't the only infected harvest of the dead. In 1985, about the time the NIH was terminating the cadaveric growth hormone program, Gayle Bourquin from Connecticut was undergoing an operation to remove a cholesteatoma, a benign tumor in the inner ear that, left untreated, may rupture the delicate bones of the inner ear and thereby destroy hearing. Removing the tumor means cutting into the skull from behind the ear. In Bourquin's case, surgeons also replaced part of her dura mater, the tough, thin outermost membrane covering the brain that had been damaged. The graft was Lyodura, a brand first introduced in 1969 by German manufacturer B. Braun Melsungen AG.

Like other dura mater suppliers around the globe, B. Braun went to medical school pathology labs to collect dura maters from corpses. But unlike U.S. companies, the German medical firm batch-processed their tissue, tossing dura maters from different corpses into a vat, and did not have the records to identify and trace the source of each graft. For disinfection, B. Braun soaked the pooled membranes with 10-percent hydrogen peroxide and then blasted them with ionizing radiation. The pathological prion protein, of course, simply shrugs off such insults. A single dura mater from a CJD victim could contaminate the entire batch.

That's apparently what went wrong with lot number 2105. Gayle Bourquin's Lyodura graft came from this lot, and she succumbed to CJD at age 28 in February 1987, 22 months after her operation. B. Braun recalled the lot (the FDA later advised surgeons to avoid lots 2000 to 2999).[21] On May 1, 1987, B. Braun revised its processing procedure, adding a one-hour soak in sodium hydroxide; in 1996, it withdrew Lyodura from the market entirely. The damage, however, was done. In the next few years, dura mater–induced CJD appeared in 17 countries, including Argentina, Australia, Canada, France, Germany, Italy, Spain, Thailand, the U.K., and two more cases in the U.S. All but a handful are linked to Lyodura.

Hardest hit of all the importing countries is Japan, which began using dura mater grafts in 1973 and spliced in more of the membrane than any other country in the world. Jun Tateishi of Kyushu University in Fukuoka, Japan, estimated that Japan patched in 20,000 grafts each year during the 1980s;[22] by comparison, the U.S. averaged about 4000 annually, and less than 10 percent were Lyodura.[23] Japanese neurosurgeons may have been enthusiastic about dura mater grafts because they saw them as bandages rather than organs. In all, between 1979 and 1991, 260,000 patients received dura mater grafts,[24] and as of June 2002, Japan had seen 88 CJD dura mater cases, nearly ten times that of second-place France. (By June 2002, 136 deaths worldwide had been attributed to CJD-infected dura mater grafts.) Because the incubation period can be quite long—it has ranged up to 18 years, the average being 6—another 35 to 40 Japanese may develop CJD this decade, according to Takeshi Sato, chair of the Japanese National CJD Surveillance Group.[25]

One Japanese woman appears to have contracted CJD from neuro-surgery done in 1989 — two years after B. Braun said it modified its sterilization protocol. It's not known if she got her graft from an old batch or if some contamination survived the new disinfection steps. Sodium hydroxide is a potent prion killer, but the chemical may not absolutely abolish it. "Because complete inactivation of the CJD agent in an intact tissue such as dura mater may not be achieved," wrote Ermias D. Belay, the CDC specialist in CJD epidemiology, "treatment with sodium hydroxide should not be regarded as a substitute for careful clinical and neuropathologic screening of donors. Even the most stringent donor screening and dura mater processing may not totally eliminate the potential for an infectious graft."[26] The emergence of mad cow disease led many nations to ban the use of cadaver brain tissue in surgery. Grafts today often come from membranes lining the thigh muscle or are synthetics derived from collagen.

The dura mater cases led to one of the most widely publicized medical lawsuits in Japan. In November 2001, courts ruled that the Japanese government, B. Braun, and the Japanese importer of Lyodura were accountable for the presence of prions in their product and for permitting its use. On March 25, 2002, the health ministry and the companies formally agreed to settle with the families of 20 victims for ¥1.2 billion (about $8.6 million).[27]

Hundreds of cases of medically transmitted CJD resulted from harvested corneas, hormones, and dura maters. There is a much more common type of harvest, however, and it doesn't come from cadavers. Live human beings donate more than 75 million units of blood annually (each unit is slightly less than a pint).[28] Can seemingly healthy prion-incubating humans pass a prion disease through blood?

Blood Safety

Over the years, isolated studies have reported transmitting the classic forms of CJD via blood transfusions to rodents. But most research has found no clear signs that it could happen. The TSE scientists at the NIH took blood samples from 13 CJD patients and inoculated or transfused them into highly susceptible primates and rodents; they could find no infection in the recipients. More persuasive than experiments, however, is simple arithmetic. There have been many donors who later developed CJD, and their blood could have contaminated a large number of pools, noted Paul Brown at a February 2002 conference on blood safety arranged by Cambridge Healthtech, a medical conference organizer.

"In spite of that arithmetic, no individual has ever been identified as having contracted CJD through blood or blood products. The combination of those two facts has persuaded most regulatory agencies all over the world that they need no longer be concerned about blood"[29]—at least from donors who later turn out to have sporadic and familial CJD. In those cases, the disease appears to begin in the brain, and by the time enough of the pathological prion protein builds up and spills out to other parts of the body, the patient is most likely to be too ill to give blood.

Variant CJD, however, is a different ball of wax. The infection starts in the periphery after the ingestion of infected meat and then works its way to the brain, perhaps via the lymph fluid—a possibility considering the infectivity present in organs of the lymphoreticular system (any tissue or organ that makes or stores immune cells, such as the appendix, spleen, thymus, tonsils, and lymph nodes). These organs are thought to have a medium degree of variant CJD infectivity but a low degree of sporadic CJD infectivity.

Somewhere along the line, the pathological prion protein PrP^{Sc} makes contact with the bloodstream, although no one has managed to detect it in the blood. Certainly in principle, PrP^{Sc} could be there, because the normal prion protein, PrP^C, exists in various blood components. Most of PrP^C molecules are found in the platelets, the clot-forming bodies that plug holes in blood vessels. To a much lesser extent, PrP^C is found in immune cells (leucocytes) and red blood cells.[30] Platelets infected with PrP^{Sc} would present a huge public-health problem, because they are pooled from different donors. Like the cadaveric pituitary hormone fiasco, one bad donation could contaminate an entire batch.

Researchers have tried to determine which parts of the blood might transmit infection in a transfusion. So far, most of the evidence is showing that "the intravenous route of infection of variant CJD is reasonably inefficient," Brown summed up, but noted that "it's still early in the game to be very secure about this."[31] Brown and co-workers in France reported that a macaque monkey experimentally given BSE developed infectivity in its blood—in particular, in the part of the blood that becomes the buffy coat after the blood is spun in a centrifuge. The buffy coat is the layer separating the heavier red blood cells from the lighter platelet-filled plasma. The scientists inoculated the buffy coat from the infected macaque's blood into the brain of a lemur, which subsequently got sick.

More compelling evidence that blood poses a vCJD risk comes from an ongoing study at the Institute for Animal Health in the U.K. In 1998, researchers extracted blood from 18 Cheviot sheep, each of which had been infected with brain homogenates from BSE cattle (5 grams shot down the throats of 17 sheep, 0.05 gram inoculated into the brain of one). Twenty-four Cheviots from scrapie-free New Zealand were subsequently given blood at different times from those sheep, which at that point were free of any prion disease symptoms. In September 2000, the team reported that sheep D505 began developing clinical signs of a prion illness 610 days after getting the transfusion. Criticism greeted this preliminary result—but then, less than two years later, a second sheep, F19, started getting sick, 538 days after receiving a transfusion from a different sheep. Both donor sheep were about halfway into their own incubation periods. (Two other sheep, which got blood from clinically ill donors rather than from asymptomatic ones, were also showing signs of prion disease by the fall of 2002.)[32]

So far, there's been no documented case of prion disease transmission in humans via blood. In the U.K., "we have a study going on where we find out if any of the variant CJD cases have been blood donors," Robert Will of the CJD Surveillance Unit explained.[33] As of early 2002, eight people who had given blood later died of vCJD. "We have a list of 22 individuals who received the blood," Will said, and so far all 22 remain vCJD-free. "Much more complicated is that some of the blood donated by variant cases went for plasma fractionation," done to collect platelets. The platelets are pooled, he added, so that "there are tens of thousands of people who were potentially exposed to pools," although the concentration of any vCJD prions would have been greatly diluted.

"Most of the transfusions took place within the last few years, since 1995, 1996," Will explained. So it would take years to know if the recipients contracted vCJD. Although animal studies are showing that transfusions can deliver a fatal prion disease, the route seems rather inefficient. Still, Will noted, "at the end of the day, regardless of how many animal studies you do, the only way you will know whether variant CJD was transmitted through blood or blood products is to actually do this look-back study" of blood recipients, which will take years to complete.

There is no test to determine the presence of malformed prion proteins that might be circulating in exposed individuals. Several companies have teamed up with university researchers and the Red Cross to develop technologies to detect the prion protein in blood and blood products. In the coming years, scientists might rely on membranes with nanometer-sized holes to filter out prion proteins. Or they may use a small molecule that can latch onto a portion of the prion protein—both molecule and prion would then be washed away. But until a validated means to screen vCJD-infected blood materializes, health officials will have to depend on various precautionary measures. The U.K. "leucodepletes" its donations—that is, it removes the white blood cells. These cells are part of the infection-fighting lymphoreticular system, which might have a concentration of infectious prion protein. For those born after 1996, when vCJD was identified and more stringent agricultural regulations came into force, the U.K. gives plasma imported from the U.S. Britain also formed plans to bar anyone who received blood from giving it.

The U.S. Food and Drug Administration, on advice of the Transmissible Spongiform Encephalopathy Advisory Committee (TSEAC), has erected blood blockades. All blood imports from Europe are banned. Donors are disqualified if they have lived in the U.K. for more than three months between 1980 and 1996, or in France for five years since 1980. The agency also tightened the donor eligibility for U.S. military personnel and their dependents who were stationed in northern Europe, and in October 2002 it extended the donor rules to include anyone who has lived in Europe for five years. (The American Red Cross, which collects nearly half of the U.S. blood supply, has generally stricter requirements: The U.K. donor ban doesn't stop at 1996 but continues to the present. The Red Cross also instituted European-wide donor bans before the FDA did.[34]) The disqualifications, called donor deferrals, cut the theoretical risk of getting variant CJD through blood by 90 percent and have resulted in about a 5-percent donor loss.[35] Taking the brunt of the blood loss is New York City, which had imported up to 35 percent of its blood from Europe, or roughly 140,000 units a year.[36]

Of course, the FDA would ban blood from anyone who has lived in Europe if the nation could afford to lose that many donors. It can't, so the numbers were picked based on pragmatic concerns. Many blood centers, which have had to expand their collection efforts, remain worried. America's Blood Centers, a national network of community blood banks responsible for most of the half of the U.S. supply the Red Cross does not collect, noted an increase in vCJD-related travel disqualifications from 0.1 percent in June 2001 to 1.4 percent in June 2002. The rules "are having a serious impact on the blood supply," the organization told the TSEAC committee during its June 2002 meeting. It asked TSEAC and the FDA about the criteria for eventually lifting some or all vCJD-related travel disqualifications.

Balancing the theoretical risk of contracting vCJD from blood and the real risk of not having enough blood in an emergency is "a very complicated situation which may get worse," Will mentioned. "If you take a lot of action, you end up doing a lot of harm." One tragic set of injuries and deaths may have resulted from just this sort of precaution. On January 4, 2001, the U.K. Department of Health ordered a switch to single-use disposable surgical instruments for tonsillectomies and adenoidectomies (single-use instruments had already been ordered for use where possible on procedures dealing with higher-risk parts of the body, such as lumbar punctures[37]). One such instrument was the electrosurgical (diathermy) forceps, which enable physicians to seal blood vessels with heat rather than with stitches.

But then came several reports of postoperative bleeding—it rose from 3 percent to 20 percent in some hospitals—and the deaths of a 33-year-old woman and a 2-year-old boy. In cauterizing blood vessels, the forceps evidently also damaged the underlying tissue so that it could not remain intact. The U.K. medical authorities ordered a return to the traditional instruments later that year.

Dental Danger

Invasive surgery may not be the only way to spread vCJD iatrogenically; trips to the dentist are not completely hazard-free. Researchers in Rome inoculated scrapie into the gut of hamsters and found that their tooth pulp and gums "bore a substantial level of infectivity"—on par with that seen in tonsils in human vCJD patients.[38] The Italian team was also able to infect other hamsters with the tooth pulp. Presumably, PrP[Sc] had made its way to the oral tissues via the trigeminal nerve, a large, three-branched facial nerve that connects to the brain stem.

Filling cavities doesn't pose a vCJD hazard, but root canals might. Researchers at the Glasgow Dental Hospital and School collected endodontic files—sharp instruments that may graze the tips of the trigeminal nerve. Light and electron microscopy revealed that 22 out of 29 files from general dental practices and 5 of 37 from dental hospitals were still visibly contaminated, even though they had undergone standard cleaning and sterilization.[39]

In reviewing the issue of prions and dentistry for the Royal Society of Medicine, Stephen R. Porter of the University of London wrote that "at present there are no data to suggest any clustering of variant CJD (vCJD) about a dental practice" but noted that dental waterlines could suck up prions. Retraction of oral fluids means that patients' bacteria could get into the tubes and form sticky biofilms that are difficult to remove. "At present the dental instruments of patients with known prion disease should be discarded after use," Porter recommended, and suction devices and waterlines should not be used on prion disease patients.[40]

Iatrogenic risks are not confined to Europe, where in addition to the U.K., Ireland, France, and Italy have seen vCJD cases. The mobility in modern civilization means that asymptomatic vCJD incubators have dispersed around the world. Hong Kong saw a case because the woman spent years in England (she was counted as a U.K. casualty). North America got its first cases in 2002. In the U.S., it was a 23-year-old Florida woman who lived in the U.K. until she was 13. In Canada, it was a Saskatchewan man who spent significant time in the U.K. Inevitably, more vCJD cases will turn up in North America.

The Florida case, "Charlene," did not have any major surgeries, and considering that the spread of human prions via the dental route is still just a theoretical possibility, there's not much concern for iatrogenic transmission originating from her. But the Saskatchewan man had an endoscopy, and that endoscope was subsequently used on 70 other people.[41] As U.K. researchers have pointed out, endoscopy on patients incubating vCJD may result in exposure of the instrument to PrP[Sc], and existing sterilization protocols may not remove all the infectious material.[42] (Sporadic CJD is not thought to pose a contamination risk during endoscopy.)

Beyond Beef

Humans may not be the only source of iatrogenic vCJD infection. BSE cows could be, too. One of the biggest concerns, especially early on in the BSE epidemic, was whether products derived from cattle could harbor infectivity—vaccines, gel caps, dietary supplements, and the like.

About 53 to 70 percent of a bovine actually turns into meat for humans. The remaining cattle parts become animal feed or find their way into an astonishing variety of nonfood products. "Indeed, it has been said, and not altogether facetiously," the U.K. BSE Inquiry noted, "that the only industry in which some part of the cow is not used is concrete production."[43] Besides soaps and candles, bovine fat turns into toothpaste, topical ointments, chewing gum, and lubricants. The small intestine makes strings for racquets and musical instruments. Fire-extinguishing foam comes from the horn, blood, and plasma. Gelatin, derived from collagen, appears in numerous products ranging from jelly candies and medicines (including soft gel capsules) to photographic chemicals. And of course the skin becomes leather.

Some medical products come directly from the cow—sutures, for instance, often derive from the intestines. Some of the cow's hormones and other proteins, such as insulin and heparin (a blood thinner) are on pharmaceutical shelves. In fact, a small population of insulin-dependent diabetics prefers the insulin harvested from cows, rather than using insulin genetically engineered by splicing the gene for insulin into bacteria and letting them make the hormone. (These diabetics say they can better detect signs of insulin shock with the bovine product.) The slaughterhouses remove the necessary organs and ship them frozen to pharmaceutical firms, which put the material through several purification steps to get the desired product.

The question is whether these products harbor enough of the BSE prions to become a danger to human health. Prion proteins are concentrated in the brain, spinal cord, and the lymphoreticular system. Worry focused on vaccines in particular—not because they contained cow parts, but because the production of the medicine often requires the use of nutrients derived from bovines. These nutrients go to feed the yeast, bacteria, or virus-infected cells that make the critical proteins that serve as the vaccines.

For virus-infected cells, a common nutrient is serum, the pale-yellow liquid component that separates from clotted blood (it's basically plasma minus the blood-clotting substance fibrinogen). In newborn or fetal calves, serum is packed with nutrients and growth factors, far more than the blood serum from adults. A needle injected into the newborn or fetal calf draws out the blood. The serum is extracted from the blood and filtered, then bottled and frozen. Another type of food for drug-making microorganisms is a beef broth. Commonly used to grow bacteria, it is based on peptone, created through the chemical treatment of milk or meat. When the cells are harvested, they are washed to remove the serum or nutrient broth. As the BSE epidemic picked up steam in the late 1980s, the U.K. sought to cut the risk from bovine-derived medicines and turned to cows from outside Britain. But still, vaccine manufacture isn't an overnight process; producing safe drugs can take years, because makers have to start entirely new seed colonies of cells. Because allowing children to go without their shots was inconceivable, existing stocks had to be used.

In December 2001, the U.K.'s Spongiform Encephalopathy Advisory Committee (SEAC) noted that two vCJD patients had gotten their oral polio vaccine from the same batch in 1994. The vaccine in this batch had been produced with fetal calf serum at a time when BSE was rampant. SEAC concluded that the connection was coincidental, although several U.K. media reported it as a possible causative link. (The British government had by then, in October 2000, recalled the vaccine since the fetal calf serum should have been obtained from a BSE-free nation.) Even if vaccines were made from BSE cows, the U.S. FDA's Center for Biologics Evaluation and Research (CBER) estimates that the chance of infection is minuscule. Combining estimates of the BSE incidence in the 1980s with the maximum chance that a calf could contract BSE from its mother (pegged at 10 percent), and factoring in various dilutions during processing, the CBER estimates that 1 in 40 billion doses theoretically could transmit vCJD—about 1 case every 5000 years.[44]

The rate, however, is not a formal risk assessment, because each factor has its own unknowns, and the overall risk compounds the uncertainties. In any case, it would be prudent to have pharmaceutical firms fulfill their cattle needs from countries not at great risk for BSE. That only makes sense, if not for health reasons then at least for public confidence. So the FDA sent letters in 1993 asking drug makers to stop using material derived from cattle from Britain and other BSE-risky countries. The urging was repeated in 1996, when the FDA strongly recommended that drug firms take "immediate and concrete steps." But because the FDA's statements constituted guidelines, not regulations (which would have taken longer to implement), evidently drug companies didn't feel obligated to follow them. During a routine review of a company's license application in 2000, the FDA discovered that the firm was using material from cattle from a high-risk nation. Soon the agency found four other drug makers doing the same thing, for a total of five companies making nine vaccines.

Most of these were not small biotech businesses making medicine for rare illnesses, either. Giants were involved: GlaxoSmithKline, American Home Products, Baxter International, Aventis Pasteur. Some of the drugs are well-known: the DTaP vaccine (diphtheria, tetanus toxoids, and acellular pertussis), as well as vaccines for haemophilus influenza B, hepatitis A and B, anthrax, and rabies.[45] Aventis used cattle blood from the Netherlands and didn't bother to seek another source because its scientists thought the disease agent couldn't survive the production process. BioPort, which made the rabies and anthrax vaccines in question, said it didn't know the FDA wanted it to change its seed cultures created before 1993. American Home Products had been working to change, but it was taking time: The firm needed 23 bacterial seed cultures to make its pneumonia vaccine but could only change one at a time.[46]

So far, vaccines pose only a theoretical risk—a very small one at that, although early in the BSE crises, safety concerns about them were certainly justified. There is, however, one area in the health industry that the FDA doesn't regulate, that packages cattle brains and other parts, and that doesn't have to list ingredients or sources: dietary supplements.

Mystery Pills

It's not hard to find at your neighborhood health store various products that purport to boost your brain power, enhance your vision, or get your sex drive going. Although they may be marketed as "herbal" supplements, they may not contain any herbs at all. In fact, the source material may be raw animal parts: for the pills that make you smart, cow brains; for vision, cow eyeballs; for sexual potency, cow testicles. Scott A. Norton, a physician from Chevy Chase, Maryland, found that supplement makers do not want to advertise that fact. One nationally distributed product, he found, contained 17 bovine organs: brain, spleen, lung, liver, pancreas, heart, kidney, intestine, lymph node, thymus, pituitary gland, placenta, and the like. Bull testicles were obscurely listed as "orchis."[47]

More distressing, the source material could be cattle from BSE nations; the USDA ban on bovine products only extends to food and medicine, and supplements don't fall under either category. The federal government has little authority in regulating the industry, thanks to the 1994 Dietary Supplement and Health Education Act. Labels often do not list the country of origin—nor are they required to do so, although the FDA recommends that the bovine material be obtained from non-BSE countries.

The CDC has investigated reports of CJD victims who consumed cow brains in the form of pills taken daily for many years. "It was a little worrisome to learn the ingredients of what she was taking," Paul Brown said of one victim. "She was taking a half a gram of brain of bovine origin, which the label said was 'imported,' as if this was a merit. It was not specified from which country it was imported. Here's a lady taking half a gram of brain for years, not knowing the origin of the brain, at a time when BSE was rampant."[48] But so far, there is no evidence that humans have contracted a prion disease from dietary supplements.

Potential BSE material can also enter the U.S. via cosmetics. Stearic acid, stearate, tallow, oleic acid, collagen, glycerin, gelatin, and tallow derivatives go into lipsticks, hair gels, shaving creams, and moisturizers. By and large, such toiletries are almost assuredly safe from BSE. "The processing of tallow derivatives and gelatin both involves steps that just massively reduce infectivity," Brown remarked. "We know that tallow doesn't have much infectivity to begin with, even when it's taken from a

BSE cow. And ditto for gelatin." Worries about gelatin sparked concern about a jelly candy called Mamba Fruit Chews in New York City in March 2001. The level of concern was so great that the city's department of health began looking for the German-made confection in various stores. (A city councilman subsequently called for better labeling rules for gelatin products.) Still, some of the more exotic anti-aging and anti-wrinkle creams are just lightly processed or simply chilled cattle-brain extracts, such as lipids from cell membranes. Prion protein is not likely to penetrate intact skin, but because these products get close to the lips and eyes, the prion protein could find an entryway into the body.

Contracting a prion disease from cosmetic products or dietary supplements is theoretical; no observational or experimental study has proven that such products can infect. Enough time has passed since the outbreak of BSE that it is safe to say that the stuff in personal care jars poses virtually no risk, since it is not ingested. But the same cannot be said for supplements and all their mystery components, which their manufacturers want you to consume every day. Ideally, all the ingredients would be spelled out, as well as their countries of origin, so that a consumer could weigh the risk with the dubious benefits of the supplements.

One man's misfortune is another man's opportunity, and nowhere is that morbidly more true than in medicine. If people are becoming sick—or if they are just afraid of becoming sick—then there might be a way to address these concerns. Certainly, people want to know that the food they eat is safe, that their cows, sheep, elk, and deer are healthy, that the blood they receive will not kill them. Or even if somehow they got infected with the pathological prion protein, it wouldn't constitute a horrible death sentence as it does today. Nothing is coming soon that meets all these needs, but scientists are getting closer to a test that works on asymptomatic prion cases and have even started clinical trials in the hopes of one day treating this most untreatable disease.

CHAPTER 13

Searching for Cures

New hope that the death sentence of prion diseases might someday be lifted.

The news sparked front-page headlines. Twenty-year-old Rachel Forber, tentatively diagnosed with variant Creutzfeldt-Jakob disease, had flown with her family to Stanley Prusiner's University of California, San Francisco lab to get an experimental treatment—one that hadn't even been tested on animals. But Rachel, who began showing signs of vCJD in December 2000, was deteriorating quickly, and under the "compassionate use" rules of the Food and Drug Administration, drugs already approved for other uses could be administered in such dire circumstances. With nothing to lose, the Forbers hoped for the best, and Rachel began swallowing quinacrine in late July of 2001. Shortly after the family returned to the U.K, her father told British reporters that Rachel had improved—she was able to get out of bed, walk on her own, even swim.

The bitter drug that the UCSF scientists had prescribed was widely used during World War II to combat malaria. Although effective against the disease, quinacrine had a tendency to poison the liver, imparting a yellow hue to the user's skin as the organ failed. In the following decades, quinacrine languished at the back of the pharmaceutical shelves while newer and safer antimalarials took over. But when Rachel Forber took the medication, she perked up, probably because the drug delivers an amphetamine-like kick to the central nervous system, not because quinacrine was wiping out the misfolded prion proteins. Late-stage CJD patients who were given the drug in the next year also experienced the same temporary cognitive improvement: One woman, for instance, suddenly began smiling at her doctor and family, and a man with fluttering eyes was able to fix his gaze. All lapsed back into their original unresponsive states in a matter of weeks. All soon died.

New Use for Old Drugs

Although word of Rachel Forber's apparent improvement fueled much false hope, the failure of the treatment came as no surprise. More than a decade earlier, Paul Brown and his pioneering National Institutes of Health colleagues had tested chloraquine, a relative of quinacrine, and found it to be ineffective. Indeed, since the 1970s, the NIH workers had tried some 60 compounds—antibacterials, antivirals, antifungals, antiparasitics, immunosuppressants, immunostimulants, hormones, and others. They deployed the compounds before, during, and after animals were infected. In a typical experiment, "when the drug was administered at the same time as the infection, some animals never became ill, and most had a significantly prolonged incubation period," Brown wrote.[1] The only persons likely to be able to take an anti-prion drug at the time of infection are lab workers who accidentally expose themselves by—as in the case of one researcher—slicing their hands with a razor that previously cut infected brain tissue. (The good news for this fellow is that others have also exposed themselves, but no one has yet contracted a prion disease from an accident in a lab.)

Other than when treatment was given at the time of infection, "the most that anything has ever done," Brown remarked, "is to prolong the incubation period."[2] He then pointed to a graph showing the effects of three different drugs. "By the time you move three weeks away in either direction [from the instance of infection], you essentially have no effect." When given at the start of symptoms, almost nothing worked. The most effective compound was MS-8209, a salt of the antifungal amphotericin B; even at almost toxic doses, MS-8209 only lengthened survival times by about 10 percent, a disappointing result because it inhibits the refolding of the normal prion protein (PrP^C) into the abnormal shape (PrP^{Sc}) in the test tube. (The result also demonstrates why care should be taken in drawing conclusions from in vitro studies.)

The emergence of variant CJD encouraged scientists to reexamine some of those old compounds and to search for new ones. Byron Caughey, Richard Race, and their NIH Rocky Mountain Labs colleagues reported several chemicals that exerted anti-prion effects—at least in cells growing in a petri dish. Typically, the cells are scrapie-infected mouse neuroblastoma cells, which are useful because, being cancer cells, they rapidly make more of themselves. Chemicals called sulfated glycans, tetrapyrroles, and even Congo red, the dye that renders prion rods visible, inhibit the formation of the PrP^{Sc}. Other researchers found compounds that cleared PrP^{Sc} out of cells, but the drugs themselves were poisonous or had to be given in toxic doses. For instance, statins, a family of cholesterol-lowering drugs, can prevent the formation of PrP^{Sc} in cell cultures, but the concentration needed is so high that the statins fatally deplete the fat that constitutes cell membranes.

Even if a chemical works well against the rogue prion protein and is nontoxic, it must be able to reach the area where normal prion proteins are being converted into their deadly counterparts—namely the brain. That's not easy. The brain is well protected against foreign substances that might be flowing in the bloodstream. The walls of capillaries that feed neurons are densely packed with cells, more so than the capillaries of other tissues. As such, they do not permit large or highly charged molecules from seeping between the spaces of the cells. (The capillary walls can respond to chemical changes, thereby permitting substances such as glucose, an important brain fuel, to pass.) This "wall" between the circulatory system and the neurons is called the blood-brain barrier, and it prevents pathogens and poisons from gaining access to the delicate neurons and supporting cells. It also limits the kinds of drugs that can reach the brain.

In the search for anti-prion compounds, Caughey, with researchers Katsumi Doh-Ura and Toru Iwaki of Kyushu University in Fukuoka, Japan, reported discovering two compounds that were extremely proficient at blocking the conversion of PrP^C to PrP^{Sc}. One was a protease inhibitor called ED-64; the other was quinacrine.

The work received no publicity—neither the NIH nor the *Journal of Virology*, which published the paper in May 2000, alerted the media.[3]

(The title of the paper, "Lysosomotropic Agents and Cysteine Protease Inhibitors Inhibit Scrapie-Associated Prion Protein Accumulation," did not help.) A year later, however, in August 2001, quinacrine made the newspapers nationwide in stories about Rachel Forber. No doubt the high-profile nature of the Prusiner laboratory, the media-savvy UCSF public affairs office, and the willingness of the Forber family to talk to reporters led to the attention.

The impetus behind the UCSF investigation, which began in early 2001, was postdoctoral fellow Carsten Korth, now at the University of Düsseldorf. A psychiatrist by training, Korth approached the problem of CJD treatment from a different angle: Rather than develop new drugs, why not look at existing drugs? "We had this approach where we looked at drugs that have already been used in treating brain diseases and therefore have been known to pass the blood-brain barrier" and might have the side effect of clearing prions, Korth recounted. "We tested 15 groups of drugs that I use in neurology and psychiatry, and only one group had this prion-inhibiting effect: phenothiazine derivatives used to treat schizophrenia."[4]

The particular anti-schizophrenia compound that was tested first, chlorpromazine, wasn't perfect—it left some PrP^{Sc} in cell cultures after a week's treatment. After Korth had shown Prusiner the results, Prusiner suggested looking for other, similar compounds. So Korth hit the books, learning that chlorpromazine was derived from methylene blue, a dye created in 1876. Scientists of that time noticed that certain dyes preferentially stained microbes and so investigated the colorings' potential in killing pathogens. In 1891, German physician Paul Ehrlich (1854–1915) discovered that methylene blue could kill the malaria parasite, and he began using it to treat victims. Methylene blue eventually spawned several other kinds of antimalarials, including quinacrine. Korth noticed that quinacrine had structural characteristics similar to chlorpromazine (namely, a three-ring, or tricyclic, scaffold with a chain of molecules extending off to the side). He tried quinacrine in cell cultures, and sure enough, it worked even better than chlorpromazine, clearing the infected mouse cells of PrP^{Sc} at one-tenth the dose.

The evidence suggests that quinacrine doesn't eliminate PrP^{Sc} on its own but enables the cells to do so. This idea is not as surprising as it sounds, because cells have to get rid of misfolded and misassembled proteins all the time. Cells' quality-control mechanisms deliver these mistakes, as well as proteins that have outlived their utility, to organelles called lysosomes and proteasomes. These intracellular garbage containers not only collect the unwanted proteins, they also chop them up with digestive juices. Prion diseases, as well as other amyloid-making illnesses such as Alzheimer's and Parkinson's diseases, may occur because cells cannot clear out the clumps of garbage protein fast enough. "Quinacrine, for whatever reason—whether it's interfering with the conversion of PrP^C to PrP^{Sc}, whether it's binding to the beta sheets of PrP^{Sc} and opening it up so it can be degraded faster"—makes PrP^{Sc} easier to get rid of, said UCSF neuropathologist Stephen DeArmond. "The bottom line is that cells can clear it."[5]

Although quinacrine-treated cells in culture dishes can clear PrP^{Sc}, the same doesn't appear to hold true for cells in live animals. "No one thinks quinacrine is going to work, and in fact it doesn't work," flatly stated Paul Brown, who was outraged by the publicity surrounding the first treatments at UCSF as well as its administration in humans before animal trials. (A second patient was treated, but, as in the case for Rachel Forber, the therapy failed.) In any case, because quinacrine was an approved drug with a long history and known side effects, researchers started human clinical trials to treat prion disease. In 2002, John Collinge of the British Medical Research Council's Prion Unit at Imperial College, London, began a three-part clinical trial, one part incorporating the standard double-blind, placebo-controlled protocols. In such a study, neither the patient nor the doctor knows whether the drug or an inert substance (placebo) is being administered. Patients started at 300 milligrams daily, but they could only remain on the therapy for two months because quinacrine began damaging the liver. Scientists are still searching for a more effective drug. By the end of 2002, Prusiner's lab alone had synthesized some 10,000 compounds based on quinacrine.

Quinacrine isn't the only compound to have demonstrated temporary cognitive improvement in CJD patients. In Germany, Markus Otto of the University of Göttingen and his colleagues have been testing flupirtine, first marketed as an analgesic but then later found to prevent apoptosis, or cell suicide. During apoptosis, a genetic program instructs cells to kill themselves when they are damaged, as a means to prevent the spread of infection to other, healthy cells. The neurons of CJD victims seem to die in this way, perhaps choked by a buildup of PrP^{Sc}.

The Göttingen team initiated a double-blind, placebo-controlled study in 1997; by the end of 2002, they had 28 CJD subjects, out of 682 candidates, who met the inclusion criteria (the patients had to pass certain cognitive tests). Fifteen were given flupirtine, thirteen a placebo. Flupirtine improved the memory and orientation of patients, but the results only lasted two to four weeks, and survival times did not increase by a statistically significant amount. Otto thought that earlier administration might improve matters, but because flupirtine exerts no anti-prion effect, it is at best a temporary means to treat the symptoms. The maker of flupirtine, Asta Medica, may ask that the compound be approved as a TSE drug.

Other drugs exist that have potent anti-prion effects but have yet to reach clinical trials. Pentosan polysulfate, which is derived from beechwood shavings, is approved for use for a bladder condition called interstitial cystitis. Belonging to a class of compounds called polyanions, it has been shown to increase the life-span of scrapie-infected rodents. In fact, Christine F. Farquhar and her colleagues at the Institute for Animal Health's Neuropathogenesis Unit in Edinburgh found that repeated administration of pentosan polysulfate completely protected mice infected with high doses of scrapie. An Italian research team led by Fabrizio Tagliavini of the National Neurological Institute in Milan conducted in vitro studies that found that tetracycline and doxycycline, widely prescribed as antibiotics, bind to a certain portion of the PrP^C and prevent its folding into PrP^{Sc}. Moreover, the drugs rendered PrP^{Sc} more sensitive to protease digestion.

The problem with pentosan polysulfate and the antibiotics is that they do not penetrate the blood–brain barrier, so they can't reach the affected neurons. One way around the barrier is to administer the drugs directly into the brain. Katsumi Doh-Ura did just that, infusing pentosan sulfate into the ventricles of the brains of mice and dogs. In 2002, Doh-Ura reported finding less accumulated PrP^{Sc} in the animals and longer incubation times, and he argued that human trials should begin. Many researchers, however, frown on going so deeply into the brain. Collinge remarked that the U.K. medical authorities contemplated such a radical procedure but deemed it far too risky. But Doh-Ura's results encouraged the families of two British vCJD patients to seek the treatment. After a judicial ruling cleared the way, the infusion procedure took place in January 2003. The clamor for the dangerous procedure, which was not expected to succeed, has made British health officials rethink their position on the infusion of drugs into the brain. Even if pentosan polysulfate proves ineffective when injected intracerebrally, the large-molecule, prion-fighting compounds (which include a chlorophyll-related class called tetrapyrroles, such as porphyrins and phthalocyanines) may work well as drugs for those recently infected by, say, surgery. Alternatively, they might be used to sterilize donated blood and other tissues suspected of harboring PrP^{Sc}.

Rather than trying to pump therapeutics into the brain, it may be possible to find ways to keep prions out. Perhaps medication could destroy the paths that PrP^{Sc} takes to the brain. Prions appear to require the immune system to reach the brain. Mice lacking certain players of the immune system—such as Peyer's patches, lymphocytes in the spleen, B cells, and follicular dendritic cells—survive longer after peripheral infection (inoculation of the scrapie agent in areas other than the brain). Some in fact resist disease entirely. The trick would be to find a way to destroy parts of the immune system without paving the way for other diseases.

Rational Thinking

No one is entirely sure how anti-prion drugs work—a statement that applies to most pharmaceuticals. Few compounds are synthesized based on molecular targets identified by basic research. "The whole history of rational drug design, based on understanding the molecular mechanisms and interactions and all this sort of stuff," is not exactly filled with success stories, remarked Byron Caughey. "What really seems to work most commonly, to find therapeutics, is a shotgun approach."[6] That is, screening tens of thousands of chemicals in the hope of finding the one that might work. (This is why Caughey and the NIH Rocky Mountain Labs generate anti-prion candidates so prolifically.) Usually, compounds are first tried in vitro. If a compound doesn't work in the test tube, it can be tossed aside. If it works, scientists test it in cultured cells and then in lab animals. Finally, the drug must pass three phases of clinical trials before the Food and Drug Administration approves it. The chance that it might actually lead to a commercially successful drug is about 1 in 10,000, estimated Charles Weissmann of the U.K. Medical Research Council's Prion Unit, who created the recombinant form of interferon, a cancer fighter.

Still, basic research has led to several molecular insights that are too irresistible to ignore as possible treatment targets. One approach is to throw a kink into the trafficking of the normal prion protein, PrP^C, inside cells, thereby preventing PrP^{Sc} from being formed and building up. Normal prion protein follows a well-regulated life within cells. Created in the endoplasmic reticulum, the PrP^C makes it way through the Golgi complex, where it is modified, and then to the surface of the cell, where it stays anchored for a while. The cell membrane pinches in around the anchored PrP^C, forming an interior bubble called an endosome. The endosome moves back into the cell, taking with it PrP^C and whatever has latched onto the protein. (PrP^C may be latching onto copper ions, which cells need to regulate function.)

Researchers are looking at several molecules that might interfere with the various points on the PrP^C conveyor belt in cells, thereby bringing the trash buildup of PrP^{Sc} to an end. Much effort has been focused on keeping PrP^C from interacting with PrP^{Sc}. Engineered antibodies and short stretches of DNA or RNA (called aptamers) can bind to certain areas of either PrP^C or PrP^{Sc} and disrupt their functions.

A particularly interesting molecular-target approach comes from Claudio Soto of Serono Pharmaceutical Research Institute in Geneva. Soto wants to prevent the normal prion protein from flattening out into unwanted beta sheets and make it refold back to the correct shape, which is dominated by coils called alpha helices. Soto and his colleagues created a small protein, or peptide, that targets a piece of the prion protein. "Not just any piece," Soto explained. "The piece of the prion protein that is mostly involved in the early mis-folding event," as well as the portion of PrP^C that interacts with PrP^{Sc}.[7] Soto targeted the amino acids between codons 110 and 125, near the key section involved in beta-sheet formation (codons 115 to 132, a range that encompasses codon 129, a critical prion disease position). Soto's team created a series of compounds of similar amino-acid sequence but with a crucial difference: They added the amino acid proline, which has a hefty molecule attached to the side. Working even better than a pea under a princess's mattress, the attached peptide acts as a bulge and prevents the prion protein from flattening out into beta sheets. Soto reported in vitro experiments in which two days of treatment with the synthetic peptide reduce the number of beta sheets in prion-infected tissue by one-fifth, with a concomitant increase in the number of alpha helices. There was also a corresponding decrease in the infectivity of the sample, as indicated by increased incubation times of 15 to 30 percent after the tissue was inoculated into lab animals.

Although Soto proved the principle behind these peptides, called beta-sheet breakers, he acknowledged that there was a long way to go before such peptides could be turned into anti-prion drugs. He needs to make smaller peptides so that they won't be quickly broken down by the immune system; he also needs to tweak the chemistry of the peptides so that they can penetrate the blood-brain barrier.

What makes beta-sheet breakers interesting is that they may have use in other diseases that result in amyloid agglomeration, such as Alzheimer's disease and type 2-diabetes. In early 2003, Serono Pharmaceutical began clinical trials with beta-sheet breakers for Alzheimer's disease. Because the proteins involved in amyloid formation are different for each disease, no single beta-sheet breaker will work for all conditions; each has to be optimized for the particular protein. (The sticky bundles in Alzheimer's disease are made from beta-amyloid proteins, which are different from prion proteins.)

Other kinds of molecules can latch onto PrP^C and prevent its conversion to PrP^{Sc}. David Peretz, working in Stanley Prusiner's lab, has led studies showing the utility of certain antibody fragments called Fabs, for fragments of antigen binding. Fabs latch onto certain portions of PrP^C called epitopes, the docking region between the antibody and its target. Once docked, the Fabs prevent PrP^{Sc} outside the cell from connecting to PrP^C, or they hinder the co-factors necessary for the conversion of PrP^C. Three weeks of treatment with the most potent Fab, called D18, cured cultured cells completely. After the Fabs were removed, the PrP^{Sc} did not return, suggesting the pathological proteins were completely wiped out. The researchers also showed that infectivity was abolished. Mice inoculated with the scrapie-infected cells all died on average within 169 days, but those injected with cells treated with Fabs lived on for the duration of the experiment (more than 350 days). Researchers at Charles Weissmann's lab achieved similar results with a monoclonal antibody, 6H4, which binds to PrP^C at the cell surface and keeps PrP^{Sc} from grabbing on. A compound with the ungainly

name of phosphatidylinositol-specific phospholipase C (PIPLC) produced similar results.

Rather than using peptides or chemicals to interact with the prion
protein, the body might be able to clear the malformed protein itself,
given a little help. Vaccines stimulate an immune response by delivering
a portion of the virus, or the killed virus itself, into the body. Immune
system cells "learn" to recognize this antigen and so, when confronted
by the actual pathogen, the cells can mount an attack against it. Prion
protein would not seem like a good candidate to stimulate an immune
reaction because the natural form occurs in most of the body's cells. In
fact, one of the hallmarks of prion diseases is that they do not initiate a
defensive response because the body interprets PrP^{Sc} as one of its own
proteins.

A prion vaccine is not totally out of the question, however, and Frank
L. Heppner, Adriano Aguzzi, and their University of Zurich colleagues
proved that in principle a completely protective prion vaccine is possible. They genetically engineered mice so that their antibodies recognized prion protein. They then inoculated prions into these mice, but
the animals remained healthy thanks to an aggressive immune response
against the prions. More importantly, the mice did not show an autoim-
mune disease as a result of their anti-prion immunity, which would have
defeated any prospect of a vaccine. Scientists have begun investigating
ways to induce such an immunity in normal, "wild-type" mice. Modified versions of prion protein may be sufficient to trigger an immune
response. In 2002, researchers at New York University showed that
such recombinant prions could extend the life of infected mice.
Aguzzi's lab also demonstrated a more potent means of breaking the
body's natural tolerance for prions by introducing recombinant PrP
coupled with either virus-like particles or antibodies.

Other compounds have shown remarkable abilities in attacking
aggregated PrP^{Sc} directly. Synthetic snowflake-like molecules, called
dendrimers, have shown such an ability both in vitro and in cell cultures. A version of the branched macromolecule, called PAMAM, created by French researchers at the Centre National de la Recherche
Scientifique (CNRS) in Toulouse, decomposed prion rods in a scrapie-
infected mouse cell within a matter of hours, with a concomitant reduction of infectivity.

SUMMARY OF TREATMENT STRATEGIES

These are some general ways to treat prion diseases:

1. Stop or at least slow the expression of the normal prion protein (PrP^C), thereby giving less raw material to make PrP^{Sc}. (Presumably, other systems can take over PrP^C's function.)
2. Block the interaction between PrP^C and the as-yet unidentified chaperone molecules that help convert it to PrP^{Sc}.
3. Prevent PrP^{Sc} that is floating around outside the cell from latching onto to PrP^C sitting on the cell surface.
4. Block the conversion of PrP^C into PrP^{Sc}, which is thought to take place after PrP^C returns to the cell's interior.
5. Refold the misfolded prion protein, restoring it to its proper conformation or enabling its destruction by cellular enzymes.
6. Block the toxic effects of accumulated PrP^{Sc}.
7. Keep PrP^{Sc} from invading the central nervous system.

Although the potential therapies are promising, the sad news is that they are not for everyone. By the time a prion disease patient shows symptoms, it is far too late for treatment. Neuronal damage is so severe that current medical technology cannot help. Attempts to create a viable CJD drug aim to treat those who are infected but not yet ill, to wipe out the malformed prions before they can attack the central nervous system. Or at least, therapies might stave off the disease so that patients can live out their natural lives. But the disease is too rare to justify more than a handful of clinical trials. What's more, the cure could be deadlier than the disease. So, should a viable drug be found, physicians will need to be sure that their patients are indeed suffering from a prion disease before the remedy is given.

Diagnosing Prion Diseases

The best diagnostic test is the bioassay—putting a piece of suspect brain tissue into a lab animal and waiting to see if it becomes sick. But such tests can take anywhere from a few months to more than a year. Scientists are still searching for more efficient methods. One such alternative is the immunological assay, a test that takes four to six hours. These assays center on a search for the protease-resistant PrP^{Sc}. Collect a sample of brain tissue and mix in a protease to dissolve the normal PrP^{C}. Then add an antibody that tags the remaining prion protein. (The two dominant forms of immunoassays are called the Western blot and the enzyme-linked immunosorbent assay, or ELISA).

Because there's money to be had in the testing for mad cows in Europe and wasted deer in the U.S., some two dozen companies have entered into the business of prion diagnostics. Leading firms include Prionics AG of Zurich; Bio-Rad Laboratories, which markets a test developed by the French Atomic Energy Commission (CEA); and Enfer Technology of Tipperary, Ireland. The firms market themselves based on their purification protocols and the antibodies they use, which affect the sensitivity and accuracy of the tests. (In July 1999, the European Commission found that the BSE tests from Prionics, Bio-Rad, and Enfer all achieved 100 percent accuracy.)

Making antibodies against prion protein requires the help of rodents genetically engineered not to express PrP. "So they have never seen prion protein in their lives," explained Man-Sun Sy, an immunologist at Case Western Reserve University. "When you inject them with a prion protein, they think it is a foreign protein. Then they begin to make antibodies like crazy. Normally, if you inject prion protein into you, your immune system will say, 'this is a self protein, I should try to avoid it.' But because these mice lack prion proteins, therefore, they say, 'this is something bad.' So they make antibodies."[8] With a contract from Prion Developmental Laboratories in Buffalo Grove, Illinois, Sy has developed a range of antibodies that bind to all the possible epitopes (target sections) along the length of the prion protein. The idea is that epitopes folded away in the coils of PrP^{C} may become exposed after PrP^{C} turns into PrP^{Sc}. Therefore, with the right antibodies, PrP^{Sc} would be preferentially tagged. Protease digestion that could dissolve PrP^{C} but leave behind PrP^{Sc} would then be unnecessary.

In October 2002, researchers at Stanley Prusiner's lab reported that they had developed a test based on the idea. Called a conformation-dependent immunoassay, the technique could be much more sensitive than tests that rely on protease digestion, according to the researchers. That's because PrP^{Sc} is partially digestible, which means that the conventional tests could underestimate the total load of malformed prions.

As good as existing tests are, they require brain tissue—which by and large rules out screening live humans and animals. (For variant CJD, tonsil biopsies are used to support a diagnosis, but they are not considered definitive.) In searching for a live, or "antemortem" (before death), test, researchers have reported many factors that seem to correlate with prion infectivity—for instance, EEG readings, 14-3-3 protein in spinal fluid, scores on mental tests, and elevated levels of certain precursors to blood components. But the only validated marker for infectivity is the protease-resistant form of the prion protein, PrP^{Sc}.

The search for a live test is one of the most pressing needs in prion science. It would better ensure the safety of the blood supply and donated organs, would enable patients not yet showing symptoms to be treated prophylactically, and would prevent the reuse of contaminated surgical tools. A live test would also enable physicians to judge objectively how patients are responding to therapy. A blood test would be particularly helpful. Researchers, however, have not yet been able to detect prions in the blood; they may exist at such low levels that no assay can yet detect them.

One way to boost the efficiency of assays is to try to increase the amount of PrP^{Sc}. Researchers have found that plasminogen, a precursor to a clot-dissolving enzyme, preferentially binds to PrP^{Sc}. Through chemical precipitation over time, the compound may pull enough prions out of the bloodstream to be detectable.

A more intriguing and potentially much more significant way to enhance the detection of prions comes from Claudio Soto's lab at Serono Pharmaceutical. His team found a means to boost the amount of PrP^{Sc} much the way the polymerase chain reaction can amplify the minuscule trace of DNA left at a crime scene. The procedure, termed protein-misfolding cyclic amplification (PMCA), relies on sound waves to break up the masses of PrP^{Sc}. As a result of the procedure, there are more PrP^{Sc} pieces to convert normal PrP^{C}. You start with a sample that harbors an undetectably small amount of PrP^{Sc}. Add PrP^{C}, which gets converted; sonicate the sample (bombard it with sound waves); then add more PrP^{C}, sonicate again, and so forth, until PrP^{Sc} can be detected. The team reported a 60-fold increase via sonication in the June 14, 2001, issue of *Nature*; Soto said that a 100-fold or 1000-fold increase is possible.

With his report, Soto achieved what many labs had failed to do. "We tried the sonication things for years," remarked Byron Caughey, who had shown before that PrP^{Sc} can convert PrP^C in vitro. But he never got the conversion to become self-sustaining, which, according to the protein-only theory of prion disease, should be possible. "We had been doing it *ad nauseam*, and the one thing [Soto] did was to mix totally crude brain preparations as a source of PrP^C." Caughey wanted to be sure to identify the source that drives the reaction that turns PrP^C into PrP^{Sc}, and "that always involves a little bit of purification or a little bit of manipulation with tissue culture cells." Soto, in contrast, "had the balls to think, well, it's going to work so well we'll be able to test it by immunoblot. We just never tried that one permutation. Are we idiots or what?" Caughey chuckled. Crude brain homogenates were the key because they probably harbored the co-factors needed to trigger the conversion of PrP^C to PrP^{Sc}.

Soto's sonic-amplification method should also make it possible to prove once and for all whether the prion theory is correct. The key proof would be turning normal prion protein into the pathogenic form in vitro and then injecting the substance into an animal to test for infectivity. Sonication should generate enough new PrP^{Sc} for such a bioassay. When I spoke with Soto in December 2002, he said that he was still working on that problem, but he thought he was getting close.

Although many researchers are trying to create a viable blood test for prion disease, an unexpected prospect turned up in June 2001—one that could, in the words of one scientist, revolutionize the way the disease is handled. Having failed to find any direct evidence of prion protein in urine of any species, scientists turned to blood as a diagnostic target. But Ruth Gabizon, who went through Prusiner's lab before taking her post at Hadassah University Hospital in Jerusalem, decided to take a second look at urine—and found a protease-resistant form of the prion protein.

Gabizon didn't find PrP^{Sc} specifically—she found a new version of PrP^{Sc}. She reasoned that prion protein, which can squeeze through the kidney's filtering mechanisms, might not be detectable in urine in either its normal or rogue forms. That would stem from the effect of urea, which tends to unfold proteins (denature them) but does not digest them. She and her colleagues collected urine from scrapie-infected hamsters, BSE cows, and CJD humans and removed the urea, thereby allowing the proteins to fold back up. Once they did, they found the new, urinary form of PrP^{Sc}, which they called $UPrP^{Sc}$. Gabizon speculated that the denaturation/renaturation of urinary prions may explain why prion diseases spread horizontally among sheep and cervids: The urinary prions get excreted, and the soil then absorbs the urea, allowing the prions to refold back into the pathogenic form. Grazing animals could then pick up the pathogenic prions.

Prions may get into the urinary tract after PrP^{Sc} spills out of the brain and into the bloodstream, in the case of inherited and sporadic TSE cases. (The kidneys concentrate waste, which may explain why urinary prions are detectable but bloodstream prions are not.) Or they may come directly from the PrP^{Sc} ingested or introduced to the body in the case of variant and iatrogenic CJD. Urinary prions seem to be somewhat infectious — some of the animals inoculated with them got sick, but not all.

A reliable urine test would also be a boon to livestock testing, which currently relies on brain tissue taken after a suspected animal has been killed. In May 2002, I went to the farm of Dick and Virginia Sisco in southern New Jersey, where they maintain a flock of certified, scrapie-free sheep. Once Linda Detwiler, the USDA veterinarian in charge of BSE surveillance in the U.S., arrived, we hopped into the pens to collect sheep urine. Gabizon, as well as Richard Rubenstein of the New York State Institute of Basic Research on Staten Island, wanted samples of urine from scrapie-free sheep to serve as controls in prion-detection experiments. Detwiler would also report on how easy — or difficult — it was to collect the urine, which would bear on whether a urine test for livestock will be practical.

Sheep, I discovered, urinate at the slightest provocation. Flap your arms in front of them, and the concrete floor gets wet. But for a practical scrapie test, urine must be collected in a more controlled manner. "Hold its nose, and usually it urinates," Detwiler explained. (Rubbing the escutcheon, an area near the udder, would do the trick for cows.) A sheep was led down a fence-post chute, where Dick Sisco, a burly man, grabbed a snout with both hands. Detwiler, crouched behind the struggling sheep, held a translucent plastic cup under the animal. Almost immediately, it filled about halfway.

Getting blood was messier; Detwiler had to feel for the jugular vein under the wool as the sheep fought against Sisco. The needle is even finer than those used on humans, so it should not be painful, but the sheep didn't seem to like it at all. Blood dripped down Detwiler's gloved hands and soaked through her blue protective jump suit and spotted her pants. She even got a fleck of blood on her nose. When it comes to sheep, urine is definitely the way to go.

Whether urinary prions are a good marker for incubating prion disease cases remains to be seen. At the moment, Gabizon is tracking the course of familial CJD cases. By the end of 2002, she found that one-third of them have the urinary prion; now, to confirm the utility of urinary prions, she has to wait for all the patients to die, which she admitted is rather depressing.[9]

Considering the need—not to mention the money to be made—a live diagnostic test for prion diseases will probably be created within the next few years. It's harder to make any predictions for treatment. The many ideas and approaches that researchers have are quite astonishing and clever, and the success of experiments in cell cultures offers much hope.

Further into the future, manipulation of the way the cell produces the natural prion protein may be possible, so that it cannot adopt its rogue form. The approach is called dominant-negative inhibition. Transgenic mice with slightly mutated prion protein, in which an amino acid on the normal PrP is replaced, remained healthy after inoculation with scrapie; mice expressing both the mutant and normal PrP survived longer than completely normal mice. One team accomplished dominant-negative inhibition by shortening the prion protein: They sliced out an amino-acid sequence common to the prion proteins of all species (an area that spans codons 114 to 121). Gene therapy techniques, which rely on vectors such as viruses to reprogram DNA, could someday perform the alteration of the PrP gene. Another transgenic possibility calls for cells to create normal prion protein as two molecules joined together. These dimers cannot undergo the beta-sheet folding when they encounter PrP^{Sc}. Genetic manipulation, however, may be better suited as means to breed prion-disease-resistant livestock than as a human therapy.

Stem cells may also be a treatment possibility in the long term. Such cells can turn into any tissue, and researchers are actively trying to exploit them. British scientists have begun experimenting with grafting such cells into the brains of mice. Preliminary results indicate that the embryonic stem cells reduce the number of neurons lost to prion infection. (Because current U.S. policy limits the use of stem cells from embryos, such medical breakthroughs are more likely to occur in other countries.)

But curing prion disease is not an economically viable proposition. "The cure for the disease will not come from a pharmaceutical company," Stanley Prusiner predicted. "It will come from a research laboratory or an academic institution funded by governments."[10] A pharmaceutical firm might get involved should one of its existing chemicals prove efficacious, or if its work would bear fruit for a more lucrative disease, such as Alzheimer's disease. So, a cure will depend on the willingness of nations to commit financial resources.

Treatment possibilities also raise thorny ethical issues. How does one measure the value of a therapy if it is not a complete cure? Should a drug be given if it lengthens a CJD's victim's life by ten percent? Twenty percent? Is it morally acceptable to keep them alive when they are brain-damaged mutes lying motionless in a hospital bed? For families inflicted by hereditary forms of the disease, such as fatal insomnia, is it wise to begin prophylactic treatment on babies and continue all the way through adulthood? After all, no one knows the effects of such extremely long-term drug administration, and the victims would otherwise remain healthy until the fifth or sixth decade of life. Is it ethical to give a CJD victim a drug that merely boosts cognitive function temporarily and only weakly so? It is disheartening to see the "improvement" of quinacrine patients, knowing the effect is short-lived—and unclear as to whether the patients themselves are happier for it.

The good news is, no one would have pondered these ethical dilemmas just a few years ago, so distant was the prospect of a cure. A successful therapy is still far off—but at least now, there are reasons for hope.

CHAPTER 14

Laying Odds

Are prion diseases more prevalent than we thought?

Researchers and government officials badly underestimated the threat that mad cow disease posed when it first appeared in Britain. They didn't think bovine spongiform encephalopathy was a zoonosis — an animal disease that can sicken people. The 1996 news that BSE could infect humans with a new form of Creutzfeldt-Jakob disease stunned the world. It also got some biomedical researchers wondering whether sporadic CJD may really be a manifestation of a zoonotic sickness. Might it be caused by the ingestion of prions, as variant CJD is?

Revisiting Sporadic CJD

It's not hard to get Terry Singeltary going. "I have my conspiracy theories," admitted the 49-year-old Texan.[1] Singeltary is probably the nation's most relentless consumer advocate when it comes to issues in prion diseases. He has helped families learn about the sickness and coordinated efforts with support groups such as CJD Voice and the CJD Foundation. He has also connected with others who are critical of the American way of handling the threat of prion diseases. Such critics include Consumers Union's Michael Hansen, journalist John Stauber, and Thomas Pringle, who used to run the voluminous www.mad-cow.org Web site. These three lend their expertise to newspaper and magazine stories about prion diseases, and they usually argue that prions represent more of a threat than people realize, and that the government has responded poorly to the dangers because it is more concerned about protecting the beef industry than people's health.

Singeltary has similar inclinations, but unlike these men, he doesn't have the professional credentials behind him. He is an 11th-grade dropout, a machinist who retired because of a neck injury sustained at work. But you might not know that from the vast stores of information in his mind and on his hard drive. Over the years, he has provided unacknowledged help to reporters around the globe, passing on files to such big-time players as *The New York Times*, *Newsweek*, and *USA Today*. His networking with journalists, activists, and concerned citizens has helped medical authorities make contact with suspected CJD victims. He has kept scientists informed with his almost daily posting of news items and research abstracts on electronic newsgroups, including the bulletin board on www.vegsource.com and the BSE-listserv run out of the University of Karlsruhe, Germany. His combative, blunt, opinionated style sometimes borders on obsessive ranting that earns praise from some officials and researchers but infuriates others — especially when he repeats his conviction that "the government has lied to us, the feed industry has lied to us — all over a buck." As evidence, Singeltary cites the USDA's testing approach, which targets downer cows and examined 19,900 of them in 2002. To him, the USDA should test 1 million cattle, because the incidence of BSE may be as low as one in a million, as it was in some European countries. That the U.S. does not, he thinks, is a sign that the government is really not interested in finding mad cowsbecause of fears of an economic disaster.

Singeltary got into the field of transmissible spongiform encephalopathy in 1997, just after his mother died of sporadic CJD. She had an especially aggressive version — the Heidenhain variant — that first causes the patient to go blind and then to deteriorate rapidly. She died just ten weeks after her symptoms began. Singeltary, who said he had watched his grandparents die of cancer, considered her death by CJD to be much, much worse: "It's something you never forget." Her uncontrollable muscle twitching became so bad "that it took three of us to hold her one time," Singeltary recalled. "She did everything but levitate in bed and spin her head." Doctors originally diagnosed Alzheimer's disease, but a postmortem neuropathological exam demanded by Singeltary revealed the true nature of her death.

Classifying a disease as "sporadic" is another way for doctors to say they don't know the cause. Normal prion proteins just turn rogue in the brain for no apparent reason. The term "sporadic" is often particularly hard for the victims' families to accept, especially when the patient was previously in robust health. Maybe it was something in the water, they wonder, or in the air, or something they ate—the same questions CJD researchers tried to answer decades ago. The names "sporadic CJD" and "variant CJD" also confuse the public and raise suspicions that U.S. authorities are hiding something when they say there have been no native variant CJD cases in the country.

Singeltary suspected an environmental cause in his mother's demise—a feeling reinforced a year later when a neighbor died of sporadic CJD. For years, the neighbor had been taking nutritional supplements that contained cow brain extracts. Researchers from the National Institutes of Health collected samples of the supplement, Singeltary recounted, and inoculated suspensions into mice. The mice remained healthy—which only means that those supplement samples tested were prion-free.

Scientists have made several attempts during the past few decades to find a connection between sporadic CJD and the environment. Often, these studies take the form of asking family members about CJD victims—their diet, occupation, medical history, hobbies, pets, and so forth—and comparing them with non-CJD subjects. Such case-control CJD studies have produced some intriguing—and sometimes contradictory—results. In 1985, Carleton Gajdusek and his NIH colleagues reported a correlation between CJD and eating a lot of roast pork, ham, hot dogs, and lamb, as well as rare meats and raw oysters.[2] Yet they also recognized that the findings were preliminary and that more studies were needed.

Following up, Robert Will of the U.K. National CJD Surveillance Unit and others pooled this data with those from two other case-control studies on CJD (one from Japan and one from the U.K.). In particular, they figured the so-called odds ratio—calculated by dividing the frequency of a possible factor in the patient group by the frequency of the factor in the control group. An odds ratio greater than 1 means that the factor may be significant. In their study, Will and his collaborators found an increase of CJD in people who have worked as health professionals (odds ratio of 1.5) and people who have had contact with cows

(1.7) and sheep (1.6). Unfortunately, those connections were not statistically significant: The numbers of pooled patients (117) and control subjects (333) were so small that the researchers felt the odds ratios needed to reach 2.5 to 8 (depending on the assumptions) before they could be deemed statistically significant. The only statistically significant correlations they found were between CJD and a family history of either CJD (19.1) or other psychotic disease (9.9), although the latter might simply be correlated because psychotic disease may be an early symptom of undiagnosed CJD.[3] In contrast with earlier findings, the team concluded that there was no association between sporadic CJD and the consumption of organ meats, including brains (0.6).

Although these case-control studies shed a certain amount of light on potential risk factors for CJD, it's impossible to draw firm conclusions. Obtaining data that produces statistically meaningful results can be difficult because of the rarity of CJD and hence the shortage of subjects. Human memory is quite fragile, too, so patients' families may not accurately recall the lifestyle and dietary habits of their loved ones over the course of a decade or more. Consequently, researchers must cope with data that probably contain significant biases. In a review paper on CJD, Joe Gibbs of the NIH and Richard T. Johnson of Johns Hopkins University concluded that "the absence of geographic differences in incidence is more convincing evidence against major dietary factors, since large populations eschew pork and some consume no meat or meat products." A CJD study of lifelong vegetarians, they proposed, could produce some interesting data.[4]

The inconclusive results of case-control studies do not completely rule out the environment as a possible cause of CJD. "Dr. Prusiner's theory does fit much of the data of spontaneous generation of [malformed] PrP somewhere in the brain," Will remarked—that is, the idea that sporadic CJD just happens by itself falls within the realm of the prion theory. Still, "it's very odd, if you look at all the forms of human prion diseases there are, all of them are transmissible in the laboratory and could be due to some sort of infectious agent."[5] One of the great difficulties, he explained, is that "given that this is a disease of an extraordinarily long incubation period, are we really confident that we can exclude childhood exposure that is transmitted from person to person, as people move around? It's difficult to be sure about that." There might a "carrier state" that leaves people healthy yet still able to infect others. If so, "you would never be able to identify what's causing the spread of the disease," concluded Will, who hasn't stopped looking for a possible environmental link. He has some preliminary data based on studies that trace CJD victims' lives well before the time symptoms began—up to 70 years; they suggest some degree of geographic clustering, but no obvious candidates for a source of infection.

A Case for Undercounting

The difficulty in establishing causal links in sporadic prion diseases—if there are any in the first place—underlines the importance of thorough surveillance. The U.K. has an active program, and when a victim of CJD is reported, one of Robert Will's colleagues visits and questions the victim's family. "No one has looked for CJD systematically in the U.S.," the NIH's Paul Brown noted. "Ever."[6] The U.S., through the Centers for Disease Control and Prevention, has generally maintained a more passive system, collecting information from death certificates from the National Center for Health Statistics. Because CJD is invariably fatal, mortality data is considered to be an effective means of tabulating cases. The CDC assessed the accuracy of such data by comparing the numbers with figures garnered through an active search in 1996: Teams covering five regions of the U.S. contacted the specialists involved and reviewed medical records for CJD cases between 1991 and 1995. Comparing the actively garnered data with the death certificate information showed that "we miss about 14 percent," said CDC epidemiologist Lawrence Schonberger. "That's improving. Doctors are becoming more knowledgeable," thanks to increased scientific and media attention given to prion diseases.[7]

The active surveillance study of 1996, however, only looked at cases in which physicians attributed the deaths to CJD. Misdiagnosed patients or patients who never saw a neurologist were not tabulated—thus CJD may be grossly underreported. Many neurological ailments share symptoms, especially early on. According to various studies, autopsies have found that CJD is misdiagnosed as other ills, such as dementia or Alzheimer's disease, 5 to 13 percent of the time. The CDC finds that around 50,000 Americans die from Alzheimer's each year (about 4 million have the disease, according to the Alzheimer's Association). Therefore, one could argue that thousands of CJD cases are being missed. (On the flip side, CJD could be mistakenly diagnosed as Alzheimer's disease or dementia, but the number of CJD patients is so small that they wouldn't dramatically skew the statistics for other neurological ills.)

In part to address the issue of misdiagnosis, CJD families have asked the CDC to place the disease on the national list of officially notifiable illnesses, which tends to include more contagious conditions such as AIDS, tuberculosis, hepatitis, and viral forms of encephalitis. Currently, only some states impose this requirement. CDC officials have discounted the utility of such an approach, arguing that it would duplicate the mortality data, which is more accurate than early diagnoses of CJD, anyway. Moreover, mandatory reporting of CJD cases does not necessarily guarantee the end to missed cases.[8]

One clue suggests that the passive system is undercounting CJD in the U.S.: racial difference. The number of black CJD victims is about 38 percent that of white victims. Rather than sporadic CJD being a one-in-a-million lottery, it's more like one-in-2.5-million for African-Americans. Access to medical care might be one reason. Schonberger recounted that the CDC had asked other countries with substantial black populations to submit CJD figures for comparison but found that the surveillance in those countries was inadequate. "We haven't been able to find any comparable literature on this issue, so it's still up in the air," Schonberger said. On the other hand, Alzheimer's disease is more common among black people than whites, with an estimated higher prevalence ranging from 14 percent to almost 100 percent, according to a February 2002 report by the Alzheimer's Association. Are some black CJD cases being misdiagnosed as Alzheimer's?

Answering critics like Terry Singeltary, who feels that the U.S. undercounts CJD, Schonberger conceded that the current surveillance system has errors but stated that most of the errors will be confined to the older population. As Schonberger pointed out, no doctor would misdiagnose a 30-year-old CJD patient as having Alzheimer's. The average age of the first 100 variant CJD victims was 29; should the epidemiology of vCJD change—if older people start coming down with it—then there would be problems. "The adequacy of our overall CJD surveillance would be greatly reduced should the proportion of older individuals affected by variant CJD substantially increase," Schonberger explained.[9]

To date, only brain autopsies can confirm CJD. To encourage the necessary neuropathological studies, in 1997 the CDC helped establish the National Prion Disease Pathology Surveillance Center at Case Western Reserve University, under the directorship of Pierluigi Gambetti. But the number of brains examined has fallen far short of the number of CJD cases in the U.S.: Gambetti's lab, which receives brains based on referrals from local physicians and families, looked at only 99 sporadic CJD cases in 2000 and 138 in 2001, when about 300 each year are expected. "I'm very unhappy with the numbers," Gambetti lamented. "European countries see 100 or 90 percent of all the cases suspected. We see 30 to 40 percent."[10]

Most families don't think about having an autopsy done (which can cost upward of $1,500 if the hospitals don't pick up the tab), and members of the support group CJD Voice have said they were too distraught to think of shipping a loved one's brain by Federal Express to Gambetti's lab. (For accurate analyses of brain tissue, the autopsy must be performed within 72 hours of death, assuming the body has been kept refrigerated.) Moreover, physicians often do not suggest an autopsy, perhaps because of liability fears should the postmortem reveal that the original diagnosis was wrong. Gambetti has been working on establishing a network that would enable postmortems to be done near where the deceased person lived and without cost to the family. He is also working on advertising the existence of his surveillance center, via meetings and letters to neurologists, pathologists, and other specialists. Gambetti is also attempting to combat what he termed "hysteria" over the potential for infection that has pathologists irrationally shunning CJD cases while they willingly conduct arguably riskier AIDS autopsies. "In order to make people aware, you have to keep informing them over and over and over," he said.

Money is the main reason why the U.S. lags behind Europe in terms of surveillance. To adequately survey the 290 U.S. million residents, "you need a lot of money," Robert Will explained. "There was a CJD meeting of families in America in which poor old Larry [Schonberger] got attacked fairly vigorously because there wasn't proper surveillance. You could only do proper surveillance if you have adequate resources.

That's the bottom line. We're very fortunate in the U.K.; we have very generous resources for CJD surveillance."

Moreover, the U.K. makes feline spongiform encephalopathy an officially notifiable disease. Domestic cats proved to be good sentinel animals because they dine on the meat not fit for human consumption—the parts more likely to harbor prion infectivity. In the U.S., FSE isn't federally notifiable. And while the USDA says it has sent educational material to private veterinarians and works with vet schools,[11] it's not clear just how many vets can spot FSE, which has never been reported in the U.S. Certainly, not many cat postmortems are done.

The only active portion of the U.S. CJD surveillance system are the follow-up investigations conducted for victims of CJD under 55 years of age. It began in 1996, when young people in the U.K. started succumbing to variant CJD. Victims under 30 years of age especially arouse interest, because such cases could indicate an infection from the environment. Except for the variant CJD case in Florida, the CDC has classified all of these more youthful cases of CJD as having either sporadic or familial origins.

One such age cluster involved the three venison eaters that the CDC tried unsuccessfully to link to the deer-and-elk borne chronic wasting disease. A second grouping occurred in 2002 in a pair of Michigan men. The two—one 26 years old, the other 28—did not know each other but lived in neighboring counties in Michigan and went to the same hospital for diagnosis.[12] The CDC's investigation turned up nothing that suggested a new form of CJD had emerged.

But the increased frequency of young CJD cases is disturbing. In the 18-year period between 1979 and 1996, the U.S. had 12 cases in patients under 30, and only one of them had the sporadic form of CJD. (The other cases resulted from heredity or from transmission via contaminated growth hormone or dura mater grafts.[13]) Between 1997 and 2001, five people under 30 died of sporadic CJD: the three venison eaters and the two Michigan patients. That represents a substantial blip of five young cases in five years, as opposed to only one case in 18 years. Physicians at the University of Michigan Health System who examined the two Michigan men concluded:

> As a result of our findings, we feel that sporadic CJD may be more common than previously thought, that it may occur in younger indi-

viduals than currently perceived, and that some cases may go undiagnosed due to insufficient testing. . . . We recommend that physicians
everywhere begin to consider CJD in rapidly progressive neurological
decline of unknown causes in people under 30 years of age, and that
brain biopsy and autopsy with genetic and prion analysis be performed
in all such cases.[14]

Pathologically, the recent bout of young casualties in the U.S. appears
to be no different from CJD already seen in America. Yet theoretically
it may have come from a new source of infection, based on an unexpected result announced in late November 2002. John Collinge of the
British Medical Research Council's Prion Unit found that not all transgenic mice infected with BSE prions developed the neuropathological
and molecular characteristics of variant CJD; some of the mice instead
generated the molecular features of sporadic CJD. Therefore, some
CJD cases classified as sporadic may have actually been caused by BSE
prions, Collinge hypothesized.[15] So far, the epidemiology of CJD in the
U.K. does not bear out that supposition—there has been no substantial
uptick in sporadic CJD as would be expected if BSE could paint more
than one pathological picture. But the preliminary study, taken at face
value, could be seen as evidence that something infectious is happening
in the cases of young, sporadic CJD victims in the U.S.

Another mouse study, reported in March 2002, fueled concern that
prion infections may be more common than previously thought.[16]
Stanley Prusiner's lab found that mice infected with mouse prions accumulated PrP^{Sc} in their skeletal muscles, mostly in those in the hind
limbs. In some mice, each gram of muscle contained some 10 million
infectious doses—on par with that in the brain in other experiments
involving intracerebral inoculation. To some CJD researchers, this finding suggested that muscle meat from cows might not be safe, after all,
and that the measures taken in Europe to protect the food supply—
banning high-risk cow parts—may not be enough.

Although this study may seem alarming, its implications are not as
sweeping as they may appear. Only a minority of results in mouse studies end up having a direct analog in humans. The skeletal muscle discovery warrants further examination, but it would be premature to alter
food policies. Prions are different for each species, and accumulation of
prions varies from species to species and from disease to disease.
Furthermore, BSE cattle muscle has failed to sicken mice in bioassay

work, suggesting that little or no infectious prions lurk there. What such findings truly reveal is that prion diseases are complicated and still mysterious, and trying to quantify the risks for human health is fraught with uncertainties.

Maverick Mayhem

Such ignorance in the face of all the progress in prion research in the past several decades has left the door open to several alternative views on the origin of these diseases. The most prominent view centers on the emergence of bovine spongiform encephalopathy. According to Mark Purdey, an organic dairy farmer from Somerset, U.K., certain pesticides had something to do with the outbreak. Purdey, who declined admission to Exeter University to study zoology and psychology in favor of running a farm, is a passionate environmentalist, drawing direction from Rachel Carson's *Silent Spring*. He relies on natural ointments to treat his cows' infections, shunning the products of the chemical industry.

So the then 29-year-old dairyman became quite upset in 1984 when an official from the Ministry of Agriculture, Fisheries and Food showed up on his farm, ordering him to use a class of insecticide called organophosphates. Britain began using this type of insecticide in the 1930s to control the warble flies that burrowed into the hides of cows and rendered the cows unmarketable. In 1960, the chemical industry formulated an organophosphate pesticide that farmers could simply pour down the backs of their cattle to kill any embedded larvae. A fast-acting variation called phosmet became the most popular choice. Purdey battled the MAFF decision that ordered him to use the chemical, and he won in court, albeit on a technicality.

As the BSE epidemic began to rage, Purdey became convinced that the application of phosmet, in conjunction with the type of feed, caused mad cow disease. His investigations turned up a substantial correlation between phosmet use and local BSE cases. John Wilesmith, the veterinary epidemiologist who had concluded that changes in feed production spread BSE, found that organophosphate use was the second-most correlated risk factor, after feed: Of 169 BSE cases in 1986, 121 cows had been treated with phosmet (all had consumed meat-and-bone meal). Purdey further pointed out that organophosphates were known neurotoxins that were able to breach the blood–brain barrier, and they produced pathologies similar to those of BSE. Cats and zoo animals that contracted prion diseases got them from the insecticide in flea collars and worm medication, he figured.

At first, Purdey claimed that the insecticide itself caused BSE. Later, as evidence for the role of prion protein accumulated, he modified his theory and suggested that the insecticide encouraged the transformation of the normal form, PrP^C, to the abnormal form, PrP^{Sc}. The insecticide depleted the animals' supply of copper, an element that PrP^C grabs onto. Without any copper around, PrP^C apparently bound to a chemically similar element, manganese. The manganese may have come from chicken manure that was used in cow feed. (Chickens were fed a lot of manganese to boost egg production.) The manganese supposedly then helped convert the prion protein into its pathogenic form.

Purdey's idea was highly intriguing and drew interest from several scientists. But the bulk of the evidence points to feed as the sole vector of BSE. Furthermore, Purdey's theory cannot explain the wealth of data concerning the epidemiology of other prion diseases (both human and animal), the transmissibility studies, and the in vitro and cell culture experiments, none of which needed organophosphates to trigger the disease or convert the prion protein. Laboratory work has indicated that organophosphate insecticides can cause cells in culture to express more prion protein on their surfaces. So in theory, the compound may have made cows more susceptible to prion disease. Scientists don't find anything technically wrong with Purdey's idea, but the lack of hard evidence persuades most that, at best, organophosphates played a minor role in the BSE outbreak, if any at all.[17]

Ironically, mainstream researchers may not dismiss the layman Purdey's thinking outright, but they do that of a fellow medical man, George A. Venters. A consultant in public health medicine at the Lanarkshire Health Authority in Hamilton, Scotland, Venters raised hackles with a paper that appeared in the prestigious *BMJ* (formerly the *British Medical Journal*). In that October 2001 article, Venters argued that variant CJD does not stem from mad cow disease at all. Instead, it was the usual sporadic form of CJD, bursting into public consciousness because of improved surveillance. He said that vCJD cases resemble Hans Creutzfeldt's one and only case, Bertha Elschker, from nearly a century ago; that certain molecular tests were flawed; and that the shape of the vCJD epidemic, which should include older people, does not correspond to the BSE curve.

Many counterarguments have been directed against Venters's points: Most neurologists agree that Bertha didn't suffer from CJD as we know the disease today; Venters also failed to cite other molecular and rodent studies that support the BSE–vCJD connection; the scope of the human toll probably depends on other factors yet to be identified and can still change. But the biggest flaw in Venters's theory is this: If vCJD is an old prion disease, then why is it predominantly hitting the U.K.? Venters has poked at some of the soft spots in mainstream prion science connecting BSE to vCJD, but his argument that vCJD represents an old disease is too weak to be credible.[18]

Menu Choices

So science still has a lot to learn about prion diseases. Does it say whether you should order a burger in London or venison in Colorado? Providing a logical answer to this seemingly simple question, unfortunately, is not possible. An accurate risk assessment requires lots of data—and that is just not available for prion diseases. It would be wonderful if science had as much epidemiological information on prions as it does on, say, cigarette smoking or radiation exposure. The numbers there are so substantial that they can be sliced and recast into many interesting (and sometimes meaningless) factoids—for instance, taking two puffs of a cigarette or watching a tube television set for a year takes a person 1.5 minutes closer to death. For the prion risks, the same kinds of statistics are not yet possible.

Certainly, the odds must not be particularly great—otherwise, more than just 132 people would have contracted vCJD from eating beef made from sick cows. One newspaper story argued that the risk was so small that it could not even be quantified. That's not quite right. True, the risk of getting a prion disease from food cannot be calculated—but that's not because the risk is so tiny. It's because science just doesn't know enough about the relevant parameters, such as the dose and individual circumstances that determine why some people get it and others don't. Stephen Churchill proved susceptible—why not his sister and parents? The odd fact is, no family has had more than one case of vCJD, even though family members are likely to have eaten similar foods for years.

The only cluster of vCJD cases occurred in the small town of Queniborough, Leicestershire, where five people died of vCJD. That outbreak occurred, officials believe, because the local butcher's knife probably became contaminated and spread BSE prions to a lot of meat. And if that's the case, then why *just* five victims, rather than many more? Nobody knows.

At this point, it's safe to say that eating beef from the U.K. or elk meat from the Rockies poses no more of a threat than many everyday activities. Statistically speaking, quitting smoking, losing a few extra pounds, or talking a daily constitutional would do more for your health than avoiding beef on fears of mad cow disease. Beef versus chicken, venison versus salmon—it boils down to an emotional choice between a minuscule risk and a zero risk.

That the risks are low today, however, doesn't mean they will stay that way. The BSE epidemic and probably the CWD spread resulted from human activity. We gave prions new hosts, and these hosts, in turn, changed the nature of the prions in unpredictable ways. The emergence of prion diseases is trying to tell us something about the way we grow our food. Nowhere is that clearer than in the cattle industry.

Man-Made Madness

You can still hear echoes of it every so often on those black-and-white re-runs: a celery-dieting Lucy, say, salivating with envy as Ricky, Fred, and Ethel tear into their steaks. Beef comes across as a wholesome luxury that a middle-class family might indulge in only occasionally. That you could get a charbroiled steak today, with salad and potatoes, for less than the price of two deli sandwiches would have astounded the Ricardos and Mertzes. Post–World War II industrialization and the rise of McDonald's, Carl Jr.'s, and other fast-food chains forever changed the way cows are reared—and the way zoonotic diseases can threaten humanity.

Today's cattle industry is a wonder of efficiency and modernization, an assembly line that moos. The animals begin life on the pasture, next to their mothers. For the next few months, the calves drink their mother's milk and graze on grass—the only time in their brief lives. Soon they are weaned, and they begin feeding on grain, mostly corn, and protein supplements. They learn to eat from a trough rather than nibble off the ground. Each will spend about six months on giant feed-lots—penned areas with food at the ready—and share its meals with upward of 100,000 fellow bovines.[19] Once a cow reaches about 1000 pounds, it is shipped off to the slaughterhouse.

The rise of the feedlot system followed capitalism's inherent mandate to increase productivity—more beef for less money. A grazing steer takes four to five years to reach slaughter weight. Today's corn-fed cattle get there in less than 18 months. Plus, all those calories fatten the cow so that its muscles, which don't have to work as much as those of grazers, develop the prized marbling that gives beef its delicious flavor.

To ensure rapid growth, cattle get extra protein from supplemental feed, sometimes derived from other animals, even though bovines are vegetarians by nature. Such supplementation triggered England's BSE epidemic, so the solution in the U.S. has been to ban the use of protein derived from ruminants. (There are, however, exceptions—namely, beef tallow and certain blood components not found to have infectivity.)

But this is a solution to a problem that doesn't have to exist, if we raised cows on green pastures and not on fecal-dusted feedlots. Indeed, mad cow disease may be the least of the unexpected headaches created by industrial agriculture. Instead of grass, cows are forced to eat corn. But a cow's rumen, its first digestive chamber, is not designed by evolution to digest all that grain. It tends to swell with gas, and its acid-weakened walls can allow bacteria to seep into the bloodstream and infect the liver. So feedlot operators must pump antibiotics into the animal to control the bloat and the bacteria. Problem solved, right? Not really: the extensive use of such drugs has been linked to the emergence of antibiotic-resistant strains of bacteria. Considering the medical costs and deaths attributed to these superbugs, the 99-cent burger isn't such a good deal, after all.

All this boost in productivity and efficiency doesn't necessarily translate into dollars for ranchers; most cows offer razor-thin margins to their owners. We as consumers are the real beneficiaries. Thanks to modern agriculture, we no longer worry about nineteenth-century problems of malnutrition and deficiency diseases. The whole process of creating a high-protein product at low prices is so seductive that many developing nations are adopting the assembly-line principles of beef production. It's a good way of creating protein without taking up as much land as grazers would need.

Of course, nutrition science has advanced so that meat doesn't have to be a part of the menu at all (although it would be hard to see how a 300-pound football lineman could earn his keep without eating meat). But with today's beef prices, especially compared with those of three generations ago, it's hard to get a more delicious protein bang-for-the-buck. And accustomed to this low price, Americans are unlikely to switch to grass-fed beef, which costs more and doesn't consistently taste as good.

Meanwhile, the unnatural lives of farm animals have created an unforeseeable chain of events. Who would have guessed that protein supplements could lead to a deadly disease transmissible to humans, which could, in turn, infect other humans? Or that moving captive animals around, a common practice in agriculture, could spread chronic wasting disease among deer—so much so that, if ignored, CWD might be able to infect almost all the deer in North America?

More than just being rare sporadic or hereditary problems, prion diseases are warning us that something is out of balance, that the excessive unnaturalness we force on livestock could be catching up with us. Maybe we can keep things the way they are and patch over the problems—tighter feed ban rules here, a set of trade regulations there. Maybe they can keep our global food economy thrumming, our minds healthy, our stomachs full. Or maybe not.

Notes

CHAPTER 1: A DEATH IN DEVIZES

1 David and Dorothy Churchill, interview by the author, Devizes, U.K., October 25, 2001. All subsequent quotes are from this interview unless otherwise noted.

2 D. Bateman *et al.*, "Sporadic Creutzfeldt-Jakob Disease in a 18-Year-Old in the U.K.," *The Lancet* 346 (1995): 1155–1156.

3 Robert G. Will *et al.*, "Infectious and Sporadic Prion Diseases," in *Prion Biology and Diseases*, ed. Stanley B. Prusiner (Cold Spring Harbor, NY: Cold Spring Harbor Laboratory Press, 1999), 465–507.

CHAPTER 2: ONE IN A MILLION

1 http://www.whonamedit.com/doctor.cfm/91.html (last accessed March 6, 2003).

2 Hans Gerhard Creutzfeldt, "Über eine eigenartige herdförmige Erkrankung des Zentralnervensystems," *Zeitschrift für die gesamte Neurologie und Psychiatrie* 57 (1920): 1–18.

3 Hans Gerhard Creutzfeldt, "On a Particular Focal Disease of the Central Nervous System (Preliminary Communication)," transl. E. P. Richardson, Jr., in *Neurological Classics in Modern Translation*, ed. David A. Rottenberg and Fred H. Hochberg (New York: Hafner Press, 1977), 98.

4 *Ibid*, 99.

5 *Ibid*.

6 http://www.whonamedit.com/doctor.cfm/738.html (last accessed March 6, 2003).

7 Hans Gerhard Creutzfeldt, *Neurological Classics in Modern Translation*, 96.

8 Alfons Jakob, "Concerning a Disorder of the Central Nervous System Clinically Resembling Multiple Sclerosis with Remarkable Anatomic Findings (Spastic Pseudosclerosis)," transl. E. P Richardson, Jr., in *Neurological Classics in*

Modern Translation, ed. David A. Rottenberg and Fred H. Hochberg (New York: Hafner Press, 1977), 116.

9 *Ibid*, 120.

10 Some researchers favored "Jakob-Creutzfeldt disease" to give due credit to Jakob, who was really the one to characterize and define the illness.

11 Robert G. Will *et al.*, "Infectious and Sporadic Prion Diseases," in *Prion Biology and Diseases*, ed. Stanley B. Prusiner (Cold Spring Harbor, NY: Cold Spring Harbor Laboratory Press, 1999), 465–507.

12 Richard T. Johnson and Clarence J. Gibbs, Jr., "Creutzfeldt-Jakob Disease and Related Transmissible Spongiform Encephalopathies," *New England Journal of Medicine* 339 (1998) 27: 1994–2004.

13 Will *et al.* "Infectious and Sporadic Prion Diseases."

14 Centers for Disease Control and Prevention, http://www.cdc.gov/nchs/fastats/lcod.htm (last accessed March 6, 2003).

15 Although it may be possible that people of African descent have a naturally lower incidence: African-Americans' mortality rate from CJD in the U.S. is 38 percent that of white Americans.

16 Will *et al.*, "Infectious and Sporadic Prion Diseases."

17 Johnson and Gibbs.

18 Robert C. Holman *et al.*, "Creutzfeldt-Jakob Disease in the United States, 1979–1994," *Emerging Infectious Diseases* 2 (1996), 4 (1996): 333–337; Robert V. Gibbons *et al.*, "Creutzfeldt-Jakob Disease in the United States: 1979–1998," *Journal of the American Medical Association* 284 (2000) 18: 2322–2323; Lawrence B. Schonberger, presentation at the Cambridge Healthtech Conference on Transmissible Spongiform Encephalopathy, Washington, DC: February 6–7, 2002.

19 Colin L. Masters, and D. Carleton Gajdusek, "The Spectrum of Creutzfeldt-Jakob Disease and the Virus-Induced Subacute Spongiform Encephalopathies," in *Recent Advances in Neuropathology 2*, ed. W. T. Smith and J. B. Cavanagh (Edinburgh, U.K.: Churchill Livingstone, 1982), 139.

20 *Ibid.*, 144; Pierluigi Gambetti, personal communication.

21 T. C. Britton *et al.*, "Sporadic Creutzfeldt-Jakob Disease in a 16-Year-Old in the U.K.," *The Lancet* 346 (1995): 1155.

22 Stephen J. DeArmond and James W. Ironsides, "Neuropathology of Prion Diseases," in *Prion Biology and Diseases*, ed. Stanley B. Prusiner (Cold Spring Harbor, NY: Cold Spring Harbor Laboratory Press, 1999), 585–652.

CHAPTER 3: THE CANNIBALS' LAUGHING DEATH

1 Shirley Lindenbaum, *Kuru Sorcery: Disease and Danger in the New Guinea Highlands* (Mountain View, CA: Mayfield Publishing Company, 1979), 20.

2 D. Carleton Gajdusek, "Kuru in the New Guinea Highlands," in *Tropical Neurology*, ed. John D. Spillane (London: Oxford University Press, 1973), 379.

3 Lindenbaum, *Kuru Sorcery*, 20.

4 Vincent D. Zigas, *Laughing Death: The Untold Story of Kuru* (Clifton, NJ: Humana Press, 1990), 16. Zigas's romanticized account of events should be taken with a grain of salt. Toward the end of his life, he was suffering from an unknown mental deterioration from which he apparently died in 1983 (there was no autopsy). According to D. Carleton Gajdusek and Shirley Lindenbaum, Zigas seems to have incorporated descriptions about kuru and the Fore written later by others into his personal account.

5 *Ibid.*, 130–132.

6 Paul Brown, interview by author Bethesda, MD, February 27, 2002. All subsequent quotes are from this interview unless otherwise noted.

7 Zigas, *Laughing Death*, 226.

8 D. Carleton Gajdusek autobiography, Nobel Foundation, http://www.nobel.se/medicine/laureates/1976/gajdusek-autobio.html (last accessed March 6, 2003).

9 *Ibid.*

10 D. Carleton Gajdusek's interest in children caught up with him on April 4, 1996, after his return to the U.S. from a conference in Geneva. "I met him at the airport," Paul Brown recalled. "When we got to his house, the FBI surrounded us like Indians around a wagon train." Pistols drawn, officers moved in to arrest the 72-year-old Gajdusek on his Maryland driveway for sexually abusing one of the boys he had brought over from Micronesia. Law enforcement authorities had suspected Gajdusek for years—not just because he had brought 56 children, mostly boys, to the U.S., but also because his journal entries from the 1960s detailed, in ways that most anthropologists considered aberrant, the ritualized homosexual practices of the islanders. He described, for instance, how boys would reach into his pocket and try to fondle his penis and how they learned to perform fellatio on adults by seven or eight years of age. Gajdusek eventually pleaded guilty to molesting the boy, who was 15 at the time of the abuse and 23 when he got the Nobelist to incriminate himself in March 1996 over a tapped telephone line. Released from prison on April 27, 1998, after serving a year of his 18-month sentence, Gajdusek immediately left the U.S. for France.

11 Judith Farquhar and D. Carleton Gajdusek (eds.), *Kuru: Early Letters and Field-Notes from the Collection of D. Carleton Gajdusek* (New York: Raven Press, 1981), 31.

12 Gajdusek autobiography, Nobel Foundation.

13 Farquahar and Gajdusek, *Kuru*, 10.

14 *Ibid.*, 8.

15 Zigas, *Laughing Death*, 231.

16 Zigas, 237.

17 D. Carleton Gajdusek, *Correspondence on the History and Original Investigations of Kuru: Smadel–Gajdusek Correspondence, 1955–1958* (Washington, DC: U.S. Department of Health, Education and Welfare, 1976), 50.

18 W. J. Hadlow, "The Scrapie–Kuru Connection: Recollections of How It Came About," in *Prion Diseases in Humans and Animals*, ed. Stanley B. Prusiner *et al.* (New York: Ellis Harwood, 1992), 43.

CHAPTER 4: CONNECTING THE HOLES

1 See, for example, Ray Bradley, "Animal Prion Diseases," in *Prion Diseases*, ed. John Collinge and Mark S. Palmer (Oxford: Oxford University Press, 1997), 89–129; Linda A. Detwiler, "Scrapie," *Scientific and Technical Review (Rev. Sci.Tech. Off. Int. Epiz.)* 11 (1990) 2: 491–537; Jean-Louis Laplanche *et al.*, "Scrapie, Chronic Wasting Disease, and Transmissible Mink Encephalopathy," in *Prion Biology and Diseases*, ed. Stanley B. Prusiner (Cold Spring Harbor, NY: Cold Spring Harbor Laboratory Press, 1999), 393–429.

2 J. G. Leopoldt, *Nützliche und auf die Erfahrung gegründete Einleitung zu der Landwirthschaft, fünf Theile* (Berlin, Germany: Christian Friedrich Günthern, 1759), 348; quoted in Paul Brown and Raymond Bradley, "1755 and All That: A Historical Primer of Transmissible Spongiform Encephalopathy," *BMJ* 317 (1998) 7174: 1688–1692.

3 W. J. Hadlow, "The Scrapie–Kuru Connection: Recollections of How It Came About," in *Prion Diseases in Humans and Animals*, ed. Stanley B. Prusiner *et al.* (New York: Ellis Harwood, 1992), 40.

4 *Ibid.*, 42.

5 *Ibid.*, 43.

6 D. Carleton Gajdusek, "Kuru and Scrapie," in *Prion Diseases in Humans and Animals*, ed. Stanley B. Prusiner *et al.* (New York: Ellis Harwood, 1992), 48.

7 Hadlow, "Scrapie–Kuru," 44.

8 *Ibid.*

9 Clarence J. Gibbs, Jr., "Spongiform Encephalopathies — Slow, Latent, and Temperate Virus Infections — in Retrospect," in *Prion Diseases in Humans and Animals*, ed. Stanley B. Prusiner *et al.* (New York: Ellis Harwood, 1992), 54–55.

10 Paul Brown, interview by the author, Bethesda, MD, February 27, 2002. All subsequent quotes are from this interview unless otherwise noted.

11 Gibbs, 56.

12 *Ibid.*, 58–59.

13 *Ibid.*, 59.

14 *Ibid.*

15 *Ibid.*

16 D. Carleton Gajdusek, "Kuru in the New Guinea Highlands," in *Tropical Neurology*, ed. John D. Spillane (London: Oxford University Press, 1973), 382.

17 *Ibid.*, 379.

18 Gibbs, "Spongiform Encephalopathies," 59.

CHAPTER 5: THE BIRTH OF THE PRION

1 Tikvah Alper, "Photo- and Radiobiology of the Scrapie Agent," in *Prion Diseases of Humans and Animals*, ed. Stanley B. Prusiner *et al.* (New York: Ellis Harwood, 1992), 31.

2 *Ibid.*

3 Gordon Hunter, "The Search for the Scrapie Agent: 1961–1981," in *Prion Diseases of Humans and Animals*, ed. Stanley B. Prusiner *et al.* (New York: Ellis Harwood, 1992), 25.

4 *Ibid.*

5 I. H. Pattison and K. M. Jones, "The Possible Nature of the Transmissible Agent of Scrapie," *Veterinary Record* 80 (1967): 2–9.

6 J. S. Griffith, "Self-Replication and Scrapie," *Nature* 215 (1967): 1043–1044.

7 Hunter, "The Search for the Scrapie Agent," 26.

8 Stanley B. Prusiner autobiography, Nobel Foundation, http://www.nobel.se/medicine/laureates/1997/prusiner-autobio.html (last accessed March 6, 2003).

9 *Ibid.*

10 Gary Taubes, "The Name of the Game Is Fame. But Is It Science? Stanley Prusiner, 'Discoverer' of Prions," *Discover* December 1986: 28–52.

11 Stanley B. Prusiner, "Prions," *Scientific American* October 1984: 46–57.

12 Taubes, "The Name of the Game Is Fame."

13 Richard Rhodes, *Deadly Feasts: The "Prion" Controversy and the Public's Health* (New York: Touchstone Books, 1998), 161.

14 *Ibid.*, 161–162.

15 Prusiner derived the name "prion" by combining key letters in "*pro*teinaceous *in*fectious particles." Word of the prion hypothesis actually appeared in February in newspapers, based on a conference presentation. *The San Francisco Chronicle* referred to it as a "new form of life."

16 Stanley B. Prusiner, "Novel Proteinaceous Infectious Particles Cause Scrapie." *Science* 216 (1982): 136–144.

17 Taubes, "The Name of the Game Is Fame." With the harsh criticism from disagreeing researchers, Taubes's bruising portrait of Prusiner showed him as a glory-seeker who undermined and belittled others' work. Some researchers thought that the pummeling was personal and unfair; others considered it accurate and deserved. Since the article's publication, Prusiner has generally shied away from journalists.

18 P. A. Merz *et al.*, "Abnormal Fibrils from Scrapie-Infected Brain," *Acta Neuropathologica* 54 (1981): 63.

19 Stanley B. Prusiner, "Development of the Prion Concept," in *Prion Biology and Diseases*, ed. Stanley B. Prusiner (Cold Spring Harbor, NY: Cold Spring Harbor Laboratory Press, 1999), 93.

20 B. Oesch *et al.*, "A Cellular Gene Encodes Scrapie PrP 27-30 Protein," *Cell* 40 (1985): 735.

21 B. Chesebro *et al.* "Identification of Scrapie Prion Protein — Specific mRNA in Scrapie-Infected and Uninfected Brain," *Nature* 315 (1985): 331.

22 Stanley B. Prusiner, "Prion Diseases," *Scientific American* January 1995: 51.

CHAPTER 6: FAMILY CURSES

1 Josef Gerstmann, "Über ein noch nicht beschriebenes Reflexphänomen bei einer Erkrankung des zerebellaren Systems" ("On a Previously Undescribed Reflex Phenomenon in a Disease of the Cerebellar System"), *Wiener Medizinische Wochenschrift* 78 (1928): 906–908.

2 Colin L. Masters, D. Carleton Gajdusek, and Clarence J. Gibbs, "Creutzfeldt-Jakob Disease Virus Isolations from the Gerstmann-Sträussler Syndrome," *Brain* 104 (1981): 559–588. They note that "the peculiar crossed-arm reflex described originally by Gerstmann (1928) was not recorded in any other case, possibly because its presence was not sought. . . ."

3 *Ibid.*, 560–561.

4 F. Seitelberger, "Eigenartige familiar-hereditäre Krankheit des Zentralnervensystems in einer niederösterreichischen Sippe" ("Peculiar Familial-Hereditary Disease of the Central Nervous System in a Family in Lower Austria"), *Wiener Klinische Wochenschrift* 74, (October 12, 1962) 41/42: 687–691.

5 Masters *et al.*, "Virus Isolations," reported successful transmission of GSS isolates in three of seven brains—those three brains, however, all had spongiform change.

6 Stanley B. Prusiner, "Prion Diseases," *Scientific American* January 1995: 51.

7 *Ibid.* Being both transmissible and inherited doesn't make prion diseases unique. Some retroviruses can be passed down through the generations.

8 Pierluigi Gambetti, interview by the author in Cleveland, OH, October 5, 2001. All subsequent quotes are from this interview unless otherwise noted.

9 D. T. Max, "To Sleep No More," *The New York Times Magazine*, May 6, 2001.

10 *Ibid.*, 76.

11 *Ibid.*

12 *Ibid.*, 77.

13 V. Manetto *et al.*, "Fatal Familial Insomnia: Clinical and Pathologic Study of Five New Cases," *Neurology* 42 (1992): 312–319.

14 Moira Bruce, BBC TWO interview on "Truth Will Out," http://www.open2.net/truthwillout/CJD/article/cjd_bruce.htm (last accessed March 6, 2003).

15 Stanley B. Prusiner, "Molecular Biology of Prion Diseases," *Science* 252 (1991): 1515–1522.

16 Moira Bruce, "Truth Will Out."

17 Michael Scott *et al.*, "Transgenetic Investigations of the Species Barrier and Prion Strains," in *Prion Biology and Diseases*, ed. Stanley B. Prusiner (Cold Spring Harbor, NY: Cold Spring Harbor Laboratory Press, 1999), 307–347.

18 Paul Brown, interview by the author, Bethesda, MD, February 27, 2002. All subsequent quotes are from this interview unless otherwise noted.

19 Richard A. Bessen *et al.*, "Non-Genetic Propagation of Strain-Specific Properties of Scrapie Prion Protein," *Nature* 375 (1995): 698–700.

CHAPTER 7: ON THE PRION PROVING GROUNDS

1 Charles Weissmann, "The Prion Connection: Now in Yeast?" *Science* 264 (1994): 528–530.

2 Byron Caughey, interview by the author, Washington, DC, February 6, 2002. All subsequent quotes are from this interview unless otherwise noted.

3 Reed B. Wickner, "[URE3] as an Altered URE2 Protein: Evidence for a Prion Analog in *Saccharomyces cervisiae*," *Science* 264 (1994): 566–569.

4 Susan Lindquist, quoted in "Researchers Show that Proteins Can Transmit Heritable Traits," Howard Hughes Medical Institute news release, January 27, 2000.

5 Liming Li and Susan Lindquist, "Creating a Protein-Based Element of Inheritance," *Science* 287 (2000): 661–664.

6 John R. Glover *et al.*, "Self-Seeded Fibers Formed by Sup35, the Protein Determinant of [PSI+], a Heritable Prion-Like Factor of *S. cerevisiae*," *Cell* 89 (1997): 811–819.

7 Shu Chen, interview by the author, Cleveland, OH, October 5, 2001.

8 John Collinge at the British Medical Research Council's Prion Unit at the Institute of Neurology in London and his colleagues suggest that there might actually be four types, rather than two.

9 Byron Caughey, "Interactions between Prion Proteins Isoforms: The Kiss of Death?" *Trends in Biochemical Sciences* 26 (2001): 235–242.

10 Karen J. Knaus *et al.*, "Crystal Structure of the Human Prion Protein Reveals a Mechanism for Oligomerization," *Nature Structural Biology* 8 (2001): 770–774.

11 Witold Surewicz, interview by the author, Cleveland, OH, October 5, 2001. All subsequent quotes are from this interview unless otherwise noted.

12 Charles Weissmann, interview by the author, London, November 1, 2001. All subsequent quotes are from this interview unless otherwise noted.

13 Byron Caughey, "Interactions Between Prion Protein Isoforms," Cambridge Healthtech Institute's Eighth Annual Blood Product Safety and Transmissible Spongiform Encephalopathies meeting, Washington, DC, February 6–7, 2002.

14 Rosie Mestel, "Putting Prions to the Test," *Science* 273 (1996): 184–189. Cold fusion is the discredited theory that a nuclear fusion reaction can occur at room temperature in simple laboratory apparatuses.

15 Robert G. Rohwer, "The Scrapie Agent: 'A Virus by Any Other Name'." *Current Topics in Microbiology and Immunology* 172 (1991): 195–232.

16 Robert Rohwer, interview by the author, Gaithersburg, MD, June 27, 2002. All subsequent quotes are from this interview unless otherwise noted.

17 Bruce Chesebro, "BSE and Prions: Uncertainties About the Agent," *Science* 279 (1998): 42–43.

18 Marcia Barinaga, "Protective Role for Prion Protein?" *Science* 278 (1997): 1404–1405.

19 David R. Brown *et al.*, "The Cellular Prion Protein Binds Copper in Vivo," *Nature* 390 (1997): 684–687.

20 David A. Harris, "The Role of Copper in the Biology of PrP," Cambridge Healthtech Institute's Eighth Annual Blood Product Safety and Transmissible Spongiform Encephalopathies meeting, Washington, DC, February 6–7, 2002.

21 S. Mouillet-Richard *et al.*, "Signal Transduction Through Prion Protein," *Science* 289 (2000): 1925–1928.

22 A. Behrens *et al.*, "Absence of the Prion Protein Homologue Doppel Causes Male Sterility," *EMBO Journal* 21 (2002): 3652–3658.

CHAPTER 8: CONSUMING FEARS

1 David Bee, "Witness Statement No. 6," evidence for *Report of the BSE Inquiry* (London: Her Majesty's Stationery Office, 2000).

2 *Ibid.*

3 Carol Richardson, in *The Early Years, 1986–88*, vol. 3 of *Report of the BSE Inquiry* (London: Her Majesty's Stationery Office, 2000), 5.

4 Bee, "Witness Statement No. 6."

5 The Pitsham Farm cases were not at first recognized as a prion disease; up until the BSE Inquiry report, it was thought that the Plurenden Manor Farms presented the first ones.

6 Gerald A. H. Wells, "Witness Statement No. 65," evidence for *Report of the BSE Inquiry* (London: Her Majesty's Stationery Office, 2000).

7 Raymond Bradley, "Memo of December 19, 1986," evidence for *Report of the BSE Inquiry* (London: Her Majesty's Stationery Office, 2000).

8 *The Early Years, 1986–88*, vol. 3 of *Report of the BSE Inquiry* (London: Her Majesty's Stationary Office, 2000), 11.

9 Wells, "Witness Statement No. 65."

10 Gerald A. H. Wells *et al.*, "A Novel Progressive Spongiform Encephalopathy in Cattle," *Veterinary Record* 121 (1987): 419–420. The sick cows from Peter Stent's farm were not included, for those cases were not yet recognized to be BSE.

11 John W. Wilesmith, "Witness Statement No. 91," evidence for *Report of the BSE Inquiry* (London: Her Majesty's Stationery Office, 2000).

12 R. H. Kimberlin, "Bovine Spongiform Encephalopathy," *Scientific and Technical Review (Rev. Sci. Tech. Off. Int. Epiz.)* 11 (1992): 360.

13 John W. Wilesmith *et al.*, "Bovine Spongiform Encephalopathy: Epidemiological Studies," *Veterinary Record* 123 (1988): 638–644.

14 John W. Wilesmith, "Bovine Spongiform Encephalopathy: A Brief Epidemiology, 1985–1991," in *Prion Diseases of Animals and Humans*, ed. Stanley B. Prusiner *et al.* (New York: Ellis Harwood, 1992), 244.

15 John W. Wilesmith *et al.*, "Bovine Spongiform Encephalopathy: Epidemiological Studies on the Origin," *Veterinary Record* 128 (1991): 199–200.

16 Edward Wyatt (Bill) Bacon, "Witness Statement No. 35," evidence for *Report of the BSE Inquiry* (London: Her Majesty's Stationery Office, 2000).

17 Memo from Neville Chandler, National Renderers' Association (U.K.), evidence for *Report of the BSE Inquiry* (London: Her Majesty's Stationery Office, 2000).

18 Wilesmith, "Bovine Spongiform Encephalopathy," 247.

19 Gabriel Horn *et al.*, *Review of the Origin of BSE, July 5, 2001* (London: Department for Food, Environment and Rural Affairs, 2001), 15.

20 A study published in 1996 put this number at 54,000, although subsequent analyses doubled the estimates of BSE between 1980 and 1996.

21 Donald Acheson, in "The Southwood Working Party, 1988–89," vol. 4 of *Report of the BSE Inquiry* (London: Her Majesty's Stationary Office, 2000), 2.

22 "The Southwood Working Party, 1988–89," vol. 4 of *Report of the BSE Inquiry* (London: Her Majesty's Stationary Office, 2002), 44.

23 *Ibid.*, 1.

24 *Ibid.*

25 Department for Environment, Food and Rural Affairs, http://www.defra.gov.uk/animalh/bse/bse-statistics/level-4-weekly-stats.html#pass (last accessed September 9, 2002).

26 Brian MacArthur, "Front-Runner in the Media Herd Sparks a Stampede of Newsmen," *The Sunday Times*, May 20, 1990.

27 J. M. Wyatt *et al.*, "Naturally Occurring Scrapie-Like Spongiform Encephalopathy in Five Domestic Cats," *Veterinary Record* 129 (1991): 233–236.

28 Mammalian meat-and-bone meal is allowed for pet food, but not from specified risk materials, BSE-suspect cows, or cows over 30 months old.

29 The U.K. Department of Health announced a new management scheme that on December 1, 2001, had Robert G. Will step down as director of the CJD Surveillance Unit, a post he had held since the unit's founding; James Ironsides took over as the first in what is now a three-year rotating directorship with Will and deputy director Richard S. G. Knight.

30 Robert G. Will, interview by the author, Edinburgh, U.K., October 31, 2001. All subsequent quotes are from this interview unless otherwise noted.

31 David Churchill, interview by the author, Devizes, U.K., October 25, 2001. All subsequent quotes are from this interview unless otherwise noted.

32 Dorothy Churchill, interview by the author, Devizes, U.K., October 25, 2001. All subsequent quotes are from this interview unless otherwise noted.

33 *Report of the BSE Inquiry* (London: Her Majesty's Stationary Office, 2000), http://www.bseinquiry.gov.uk (last accessed March 6, 2003).

34 Andy Coghlan, "How It All Went So Horribly Wrong," *New Scientist*, November 4, 2000.

CHAPTER 9: MAD COW'S HUMAN TOLL

1 Food Standards Agency, *Review of BSE Controls*, December 2000, http://www.food-standards.gov.uk/bse/what/about/report (last accessed March 6, 2003).

2 Department for Environment, Food and Rural Affairs, http://www.defra.gov.uk/animalh/bse/bse-statistics/bse/general.html (last accessed January 14, 2003).

3 Robert G. Will, interview by the author, Edinburgh, U.K., October 31, 2001. All subsequent quotes are from this interview unless otherwise noted.

4 Roy M. Anderson *et al.*, "Transmission Dynamics and Epidemiology of BSE in British Cattle," *Nature* 382 (1996): 779–788.

5 Michael D. Spencer *et al.*, "First Hundred Cases of Variant Creutzfeldt-Jakob Disease: Retrospective Case Note Review of Early Psychiatric and Neurological Features," *BMJ* 324 (2002): 1479–1482.

6 Stephen DeArmond, interview by the author, San Francisco, CA, November 16, 2001.

7 Neil Ferguson, interview by the author via e-mail, March 7–8, 2001. See also Azra C. Ghani *et al.*, "Predicted vCJD Mortality in Great Britain," *Nature* 406 (2000): 583–584.

8 N. M. Ferguson *et al.*, "Estimating the Human Health Risk from Possible BSE Infection of the British Sheep Flock," *Nature* 415 (2002): 420–424.

9 Jerome N. Huillard d'Aignaux *et al.*, "Predictability of the U.K. Variant Creutzfeldt-Jakob Disease Epidemic," *Science* 294 (2001): 1729–1731.

10 Risk Solutions, "An Investigation of the Substitution of Scrapie Brain Pool Samples: A Report for DEFRA," November 2001. See also Andy Coghlan, "The Great Brain Blunder," *New Scientist*, October 27, 2001; Declan Butler, "Brain Mix-Up Leaves BSE Research in Turmoil," *Nature* 413 (2001): 760.

11 Hillary Pickles, *The Early Years, 1986–88*, vol. 3 of *Report of the BSE Inquiry* (London: Her Majesty's Stationery Office, 2000), 181.

12 Anthony Barnett, "Feed Banned in Britain Dumped on the Third World," *The Observer*, October 29, 2000, http://www.observer.co.uk/Print/ 0,3858,4083121,0.html (last accessed March 6, 2003).

13 "The Early Years, 1986–88," vol. 3 of *Report of the BSE Inquiry* (London: Her Majesty's Stationary Office, 2000).

14 "BSE-Contaminated Feed Said to Reach 70 Countries." *UPI*, February 4, 2001.

15 Tim Larimer, "A Whole Lot at Steak," *Time (Asia)*, October 14, 2001; "Tokyo Suppressed Mad Cow Risk," Associated Press, December 12, 2001.

16 James Brooke, "Mad Cow Disease Sets Off a Scare in Japan," *The New York Times*, September 27, 2001.

17 Larimer, "A Whole Lot of Steak," 2001.

18 Ray Bradley and Herbert Budka, "Scientific Report on Stunning Methods and BSE Risks," TSE-BSE Ad Hoc Group meeting of December 13, 2001, European Commission, 24.

19 Geographical BSE risk assessments by the Scientific Steering Committee can be found at http://europa.eu.int/comm/food/fs/sc/ssc/outcome_en.html (last accessed March 6, 2003).

CHAPTER 10: KEEPING THE MADNESS OUT

1 General Accounting Office, "Mad Cow Disease: Improvements in the Animal Feed Ban and Other Regulatory Areas Would Strengthen U.S. Prevention Efforts," GAO-02-183 (Washington, DC: January 2002), 14.

2 Linda A. Detwiler, interview by the author, Robbinsville, NJ, March 28, 2002. All subsequent quotes are from this interview unless otherwise noted.

3 Tim Large, "Japan Estimates Mad Cow Damage at over $2.76 Bln," Reuters, April 5, 2002.

4 General Accounting Office, "Mad Cow Disease," 32.

5 Michael Hansen, telephone interview by the author, April 26, 2002. All subsequent quotes are from this interview unless otherwise noted.

6 OIE figures, http://www.oie.int/eng/info/en_esbincidence.htm (last accessed March 22, 2002).

7 Paul Brown, interview by the author, Bethesda, MD, February 27, 2002. All subsequent quotes are from this interview unless otherwise noted.

8 See, for instance, Joshua T. Cohen *et al.*, "Evaluation of the Potential for Bovine Spongiform Encephalopathy in the United States" (Harvard Center for Risk Analysis study of BSE), November 26, 2001: 43. Available at http://www.aphis.usda.gov/lpa/issues/bse/bse-riskassmt.html.

9 "USDA & Harvard Announce Results of BSE Risk Assessment," press conference, Washington, DC, November 30, 2001, http://www.usda.gov/news/releases/2001/11/0243.htm (last accessed March 6, 2003).

10 Cohen *et al.*, "Evaluation of the Potential for Bovine Spongiform Encephalopathy in the United States," iv.

11 "USDA & Harvard Announce Results of BSE Risk Assessment."

12 Steve Stecklow, "The U.S. May Face Mad-Cow Exposure Despite Assurances from Government," *The Wall Street Journal*, November 28, 2001.

13 General Accounting Office, "Mad Cow Disease," 19.

14 *Ibid.*, 20.

15 Stecklow, "Mad-Cow Exposure."

16 Agricultural bioterrorists could probably smuggle in BSE-tainted material and cause panic if they infected various feed mills. As a means to sicken cattle, however, BSE isn't a likely weapon of choice because it is harder to transmit and much less virulent than other bovine diseases, such as foot-and-mouth disease or brucellosis. Nations that have conducted research in agricultural bioterrorism have focused on infecting crop plants; the United Nations has cited ten plant diseases that could be turned into weapons, including wheat rust, sugarcane smut, and rice blast.

17 Stecklow, "Mad Cow Exposure."

18 "Ruminant Feed (BSE) Enforcement Activities," FDA Center for Veterinary Medicine, Washington, DC, April 15, 2002, http://www.fda.gov/cvm/index/updates/bseap02.htm (last accessed March 6, 2003).

19 R. F. Marsh and W. J. Hadlow, "Transmissible Mink Encephalopathy," *Scientific and Technical Review (Rev. Sci. Tech. Off. Int. Epiz.)* 11 (June 1992), 543.

20 *Ibid.*, 544.

21 Sheldon Rampton and John Stauber. *Mad Cow U.S.A.: Could the Nightmare Happen Here?* (Monroe, ME: Common Courage Press, 1997), 89.

22 Marsh and Hadlow, "Transmissable Mink Encephalopathy," 547.

23 Mark Johnson and John Fauber, "Human Prints on Animal Crisis," *Milwaukee Journal Sentinel*, November 2, 2002.

24 Thomas Pringle, head of the Sperling Medical Foundation in Portland, OR, and former Web master of the voluminous www.mad-cow.org Web site, referred to the unit as the BSE Swat Team when he described the report after obtaining it via the Freedom of Information Act (the report is posted on the USDA's Web site). Subsequently, the team created T-shirts that actually say "BSE SWAT TEAM" on the back.

25 Department of Agriculture, "Bovine Spongiform Encephalopathy (BSE) Response Plan Summary," (Washington, DC, October 1998).

26 R. Race *et al.*, "Long-Term Subclinical Carrier State Precedes Scrapie Replication and Adaptation in a Resistant Species: Analogies to Bovine Spongiform Encephalopathy and Variant Creutzfeldt-Jakob Disease in Humans," *Journal of Virology* 75 (2001): 10106–10112.

27 Michael Hansen, Letter to Thomas Billy, Administrator, FSIS, May 15, 1997, as supplied by Hansen.

28 "Study Examines How Prion Disease Adapts to New Species," National Institute of Health News Release, October 17, 2001.

29 Dan Glickman, Secretary of Agriculture, "Declaration of Emergency Because of Scrapie in the United States," *Federal Register* 65 (2000): 14521.

CHAPTER 11: SCOURGE OF THE CERVIDS

1 Thomas Givnish, telephone interview by the author, January 27, 2003.

2 Chronic wasting disease's observed virulence may seem greater than scrapie's because certain genotypes of sheep are resistant to scrapie, whereas deer and elk seem uniformly susceptible to CWD regardless of genotype.

3 Elizabeth S. Williams and M. W. Miller, "Chronic Wasting Disease in Deer and Elk in North America," *Scientific and Technical Review* (*Rev. Sci. Tech. Off. Int. Epiz.*) 21 (2002): 305–316; Elizabeth S. Williams and S. Young," Spongiform Encephalopathies in Cervidae," *Scientific and Technical Review* (*Rev. Sci. Tech. Off. Int. Epiz.*) 11 (1992): 551–567.

4 Elizabeth S. Williams, telephone interview by the author, November 8, 2002. All subsequent quotes are from this interview unless otherwise noted.

5 Elizabeth S. Williams and S. Young, "Chronic Wasting Disease of Captive Mule Deer: A Spongiform Encephalopathy," *Journal of Wildlife Diseases* 16 (1980): 89–98.

6 Williams and Young, "Cervidae," 564.

7 Antonio Regalado, "Spreading 'Mad Deer' Plague Leaves U.S. Scientists Baffled," *The Wall Street Journal*, May 24, 2002.

8 Michael W. Miller, "Testimony Before the Subcommittee on Forests and Forest Health and the Subcommittee on Fisheries Conservation, Wildlife, and Oceans, Committee on Resources," U.S. House of Representatives, Washington, DC, May 15, 2002, http://wildlife.state.co.us/CWD/CWD_testimony51502.htm (last accessed March 6, 2003).

9 Michael W. Miller, telephone interview by the author, February 28, 2003.

10 Miller, Testimony before the Subcommittee on Forests and Forest Health.

11 The jingle can be found at http://www.bananasatlarge.com/cwd.htm, lyrics at http://www.jsonline.com/news/state/oct02/90779.asp (last accessed March 6, 2003).

12 G. J. Raymond *et al.*, "Evidence of a Molecular Barrier Limiting Susceptibility of Humans, Cattle, and Sheep to Chronic Wasting Disease," *The EMBO Journal* 19 (2000): 4425–4430.

13 Ermias D. Belay *et al.*, "Creutzfeldt-Jakob Disease in Unusually Young Patients Who Consumed Venison," *Archives of Neurology* 58 (2001): 1673–1678.

14 "Test for Chronic Wasting Disease in Humans Turns Up Negative," Associated Press, November 21, 2002.

15 Jodi Wilgoren, "Hearts Heavy, Hunters Stalk Ailing Deer," *The New York Times*, September 14, 2002; also Andrew Revkin, "Out of Control, Deer Send Ecosystem Into Chaos," *The New York Times*, November 12, 2002.

16 Mark Johnson and John Fauber, "Human Prints on Animal Crisis," *Milwaukee Journal Sentinel*, November 2, 2002.

17 Amir Hamir, telephone interview by the author, February 27, 2003.

18 Janice M. Miller, USDA, posting to BSE-listserv, November 25, 2002.

CHAPTER 12: MISADVENTURES IN MEDICINE

1 Cynthia Cowdrey, interview by the author, San Francisco, CA, November 16, 2001.

2 Shu Chen, interview by the author, Cleveland, OH, October 5, 2001. All subsequent quotes are from this interview unless otherwise noted.

3 Byron Caughey, interview by the author, Washington, DC, February 6, 2001.

4 R. D. Traub *et al.*, "Precautions in Conducting Biopsies and Autopsies on Patients with Presenile Dementia," *Journal of Neurosurgery* 41 (1974): 394–395.

5 Paul Brown, interview by the author, Bethesda, MD, February 27, 2002. All subsequent quotes are from this interview unless otherwise noted.

6 C. Bernoulli *et al.*, "Danger of Accidental Person-to-Person Transmission of Creutzfeldt-Jakob Disease by Surgery," *The Lancet* 1 (February 26, 1977): 478–479.

7 Richard T. Johnson and Clarence J. Gibbs, Jr., "Creutzfeldt-Jakob Disease and Related Transmissible Spongiform Encephalopathies," *New England Journal of Medicine* 339 (1998): 1994–2004.

8 Paul Brown, "Environmental Causes of Human Spongiform Encephalopathy," in *Methods in Molecular Medicine: Prion Diseases*, ed. H. Baker and R. M. Ridley (Totowa, NJ: Humana Press, 1996), 139–154.

9 Paul Brown *et al.*, "Iatrogenic Creutzfeldt-Jakob Disease at the Millennium," *Neurology* 55 (2000): 1075–1081.

10 Johnson and Gibbs, "Creutzfeldt-Jakob Disease."

11 Charles Weissmann, interview by the author, London, November 1, 2001. All subsequent quotes are from this interview unless otherwise noted.

12 "Inquiry Ordered into CJD Blunder," BBC News, October 30, 2002, http://news.bbc.co.uk/1/hi/health/2374107.stm (last accessed March 6, 2003).

13 Michael A. Fuoco, "Creutzfeldt-Jakob Death Prompts Letters to UPMC Neurosurgery Patients," *Pittsburgh Post-Gazette*, June 13, 2002.

14 David Sell, "Growth Hormone Is Tempting to Athletes," *The Los Angeles Times*, August 17, 1986.

15 Brown, "Environmental Causes," 141.

16 *Ibid.*

17 Emily Green, "A Wonder Drug That Carried the Seeds of Death," *The Los Angeles Times*, May 21, 2000.

18 Ermias D. Belay, "Transmissible Spongiform Encephalopathies in Humans," *Annual Review of Microbiology* 53 (1999): 283–314; Paul Brown *et al.*, "'Friendly Fire' in Medicine. Hormones, Homografts and Creutzfeldt-Jakob Disease," *The Lancet* 340 (1992): 24–27.

19 E. A. Croes *et al.*, "Creutzfeldt-Jakob Disease 38 Years after Diagnostic Use of Human Growth Hormone," *Journal of Neurology, Neurosurgery and Psychiatry* 72 (2002): 792–793.

20 Number of hormone and dura mater CJD cases from Paul Brown, presentation at the Meeting of the Transmissible Spongiform Encephalopathy Advisory Committee, Food and Drug Administration, Gaithersburg, MD, June 26–27, 2002.

21 "CJD-Linked Product Banned in U.S. Was Imported to Japan," *Japan Economic Newswire*, September 30, 2000.

22 Yosikazu Nakamura, "Relative Risk of Creutzfeldt-Jakob Disease with Cadaveric Dura Transplantation in Japan," *Neurology* 53 (1999): 218–220.

23 Brown, "Iatrogenic Creutzfeldt-Jakob Disease at the Millennium."

24 Nakamura, "Relative Risk."

25 *Morbidity and Mortality Weekly Report* 46 (1997): 1066–1069; Sonni Efron, "A Tragic Time Bomb in Japan," *The Los Angeles Times*, April 3, 2000.

26 Belay, "Transmissible Spongiform Encephalopathies in Humans,"

27 Masayuki Kitano, "Japan Brain-Wasting Disease Case Nearly Settled," Reuters, February 27, 2002; "State, Plaintiffs Formally Reach CJD Settlement," *Japan Economic Newswire*, March 25, 2002.

28 International Federation of Red Cross and Red Crescent Societies, http://www.ifrc.org/what/health/blood/index.asp (last accessed March 6, 2003).

29 Paul Brown, "Drug Therapy of CJD," Presentation at the Cambridge Healthtech Institute's Eighth Annual Transmissible Spongiform Encephalopathies meeting, Washington, DC, February 6–7, 2002.

30 Ian MacGregor, "Sources and Properties of Prion Protein (PrP^C) in Human Plasma,"Presentation at the Cambridge Healthtech Institute's Eighth Annual Transmissible Spongiform Encephalopathies meeting, Washington, DC, February 6–7, 2002.

31 Brown, "Drug Therapy."

32 Nora Hunter *et al.*, "Transmission of Prion Diseases by Blood Transfusion," *Journal of General Virology* 83 (2002): 2897–2905.

33 Robert G. Will, interview by the author, Edinburgh, U.K., October 31, 2001. All subsequent quotes are from this interview unless otherwise noted.

34 Peter L. Page, "American Red Cross, vCJD Donor Deferral Criteria, and Blood Supply," presentation at the meeting of the Transmissible Spongiform Encephalopathy Advisory Committee, Food and Drug Administration, Gaithersburg, MD, June 26–27, 2002.

35 Dorothy Scott, "Update on Implementation of Revised Guidance on Blood Donor Deferrals for Risk of CJD and vCJD," presentation at the meeting of the Transmissible Spongiform Encephalopathy Advisory Committee, Food and Drug Administration, Gaithersburg, MD, June 26–27, 2002.

36 John Tagliabue, "U.S. Plan to Halt Imports Worries Europe," *The New York Times*, July 17, 2001.

37 National Health Service (U.K.), Health Service Circular, August 13, 1999; reviewed August 13, 2002.

38 Loredana Ingrosso *et al.*, "Transmission of the 263K Scrapie Strain by the Dental Route," *Journal of General Virology* 80 (1999): 3043–3047. Toxicologists define a lethal dose (LD) as the dose needed to kill 50 percent of test animals (LD 50); it therefore corresponds to the median lethal dose. These researchers estimated gingival tissue to have 7.2 log LD 50 units (about 16 million lethal doses) per gram of tissue, and dental pulp to have 5.6 log LD 50 units (400,000 lethal doses). In vCJD, lymphoreticular systems may have 4 to 7 log LD 50 units.

39 A. Smith *et al.*, "Contaminated Dental Instruments," *Journal of Hospital Infection* 51 (2002): 233–235.

40 Stephen R. Porter, "Prions and Dentistry," *Journal of the Royal Society of Medicine* 95 (2002): 178–181.

41 Kanina Holmes, "First Canadian Dies of Human Mad Cow Strain," Reuters, August 8, 2002.

42 M. G. Bramble and J. W. Ironsides, "Creutzfeldt-Jakob Disease: Implications for Gastroenterology," *Gut* 50 (2002): 888–890.

43 Vol. 16 (Reference Material) of the *Report of the BSE Inquiry* (London: Her Majesty's Stationary Office, 2000), 66.

44 FDA Center for Biologics and Evaluation Research, http://www.fda.gov/cber/bse/risk.htm (last accessed March 6, 2003).

45 FDA, http://www.fda.gov/cber/bse/bse.htm#list, last accessed March 6, 2003.

46 Melody Petersen and Greg Winter, "5 Drug Makers Use Material with Possible Mad Cow Link," *The New York Times*, February 5, 2001.

47 Scott A. Norton, "Raw Animal Tissues and Dietary Supplements," *New England Journal of Medicine* 343 (2000): 304–305.

48 Paul Brown, as quoted in Anita Manning, "U.S. Supplements May Harbor Mad Cow Diseases," *USA Today*, January 22, 2001.

CHAPTER 13: SEARCHING FOR CURES

1 Paul Brown, "Drug Therapy in Human and Experimental Transmissible Spongiform Encephalopathy," *Neurology* 58 (2002): 1720–1725.

2 Paul Brown, interview by the author, Bethesda, MD, February 27, 2002. All subsequent quotes are from this interview unless otherwise noted.

3 Katsumi Doh-Ura *et al.*, "Lysosomotropic Agents and Cysteine Protease Inhibitors Inhibit Scrapie-Associated Prion Protein Accumulation," *Journal of Virology* 74 (2000): 4894–4897.

4 Carsten Korth, "Tricyclic Compounds and Related Molecules for Prion Therapeutics," presentation at the New Perspectives for Prion Therapeutics meeting, Paris, December 1–3, 2002.

5 Stephen DeArmond, interview by the author, San Francisco, CA, November 16, 2001.

6 Byron Caughey, interview by the author, Washington, DC, February 6, 2002. All subsequent quotes are from this interview unless otherwise noted.

7 Claudio Soto, "Beta-Sheet Breaker Peptides for TSE Therapy," presentation at the New Perspectives for Prion Therapeutics meeting, Paris, December 1–3, 2002.

8 Man-Sun Sy, interview by the author, Cleveland, OH, October 5, 2001.

9 Ruth Gabizon, interview by the author, Paris, December 3, 2002.

10 Stanley B. Prusiner, response to author's question at a press conference, Paris, December 3, 2002.

CHAPTER 14: LAYING ODDS

1 Terry Singeltary, telephone interview by the author, November 19, 2002. All subsequent quotes are from this interview unless otherwise noted.

2 Zoreh Davanipour *et al.*, "A Case-Control Study of Creutzfeldt-Jakob Disease: Dietary Risk Factors," *American Journal of Epidemiology* 122 (1985): 443–451.

3 D. P. W. M. Wientjens *et al.*, "Risk Factors for Creutzfeldt-Jakob Disease: A Reanalysis of Case-Control Studies," *Neurology* 46 (1996): 1287–1291.

4 Richard T. Johnson and Clarence J. Gibbs, Jr., "Creutzfeldt-Jakob Disease (Letter)," *New England Journal of Medicine* 340 (1999): 1688.

5 Robert G. Will, interview by the author, Edinburgh, U.K., October 31, 2002. All subsequent quotes are from this interview unless otherwise noted.

6 Paul Brown, interview by the author, Bethesda, MD, February 27, 2002. All subsequent quotes are from this interview unless otherwise noted.

7 Lawrence Schonberger, interview by the author, Washington, DC, February 6, 2002. All subsequent quotes are from this interview unless otherwise noted.

8 Robert V. Gibbons *et al.*, "Diagnosis and Reporting of Creutzfeldt-Jakob Disease (Letter)," *Journal of the American Medical Association* 285 (2001): 733.

9 Presentation at the Cambridge Healthtech Institute's Eighth Annual Transmissible Spongiform Encephalopathy conference, Washington, DC, February 6–7, 2002.

10 Pierluigi Gambetti, interview by the author, Cleveland, OH, October 5, 2001.

11 Linda A. Detwiler, interview by the author, Robbinsville, NJ, March 28, 2002.

12 The 26-year-old and Rachel Forber from the U.K. were the two patients given quinacrine as an experimental treatment in 2001.

13 Ermias D. Belay *et al.*, "Creutzfeldt-Jakob Disease in Unusually Young Patients Who Consumed Venison," *Archives of Neurology* 58 (2001): 1673–1678.

14 "Statement on Two Unusual Creutzfeldt-Jakob Disease Cases at the University of Michigan Health System," April 29, 2002, press release, http://www.med.umich.edu/opm/newspage/2002/cjd.htm (last accessed March 6, 2003).

15 Emmanuel A. Asante *et al.* "BSE Prions Propagate as Either Variant CJD-Like or Sporadic CJD-Like Prion Strains in Transgenic Mice Expressing Human Prion Protein," *EMBO Journal* 21 (2002): 6358–6366.

16 Patrick J. Bosque *et al.*, "Prions in Skeletal Muscle," *Proceedings of the National Academy of Sciences USA* 99 (2002): 3812–3817.

17 For more information on Mark Purdey's ideas, see his Web site: http://www.purdeyenvironment.com (last accessed March 6, 2003).

18 Interested readers can follow scientific thrust-and-parry between Venters and other scientists at the BMJ Web site: http://bmj.com/cgi/eletters/323/7317/858 (last accessed March 6, 2003).

19 Michael Pollan, "Power Steer," *The New York Times Magazine*, March 31, 2002.

Glossary

Cross-references are indicated in *italics*.

alpha helix: A structural shape within a protein, after the protein has folded up. It resembles a coil. (Plural: helices.)

amino acid: Basic building block of proteins. Humans rely on 20 different kinds of amino acids to produce all the needed proteins.

amyloid plaque: A starchlike substance made of protein and sugar molecules that are deposited in the brain and other organs under abnormal conditions.

antimalarial: A drug that combats malaria, such as quinacrine.

antibodies: Immune system substances that are produced in response to an infection; their function is to destroy antigens.

antigens: Proteins that trigger an immune (antibody) response.

APHIS: Animal and Plant Health Inspection Service, an arm of the USDA in charge of maintaining the health of livestock.

astrocytes: Star-shaped cells in the brain that support and nourish neurons. It is a type of glial cell. Proliferation of astrocytes (astrocytosis) happens when nearby neurons are damaged.

ataxia: Inability to coordinate the movement of voluntary muscles.

autoclave: A vessel that sterilizes medical equipment via steam under pressure; to sterilize using an autoclave.

base: One of the four coding compounds of *DNA* (adenine, thymine, cytosine, and guanine). In *RNA*, uracil substitutes for thymine.

beta sheet: A structural shape within a protein, after the protein has folded. It is a pleated, mostly flattened-out area, like a corrugated tin roof.

bioassay: Testing a sample on a live animal.

bronchopneumonia: Inflammation of the lungs; the bronchial tubes typically become clogged with mucus. Often the immediate cause of death in *prion*-disease patients because of their immobility in the late stages.

BSE: Bovine spongiform encephalopathy, or "mad cow disease"; a *prion* disease of cattle in the U.K., marked by aggressiveness and *ataxia*.

cell culture: Cultivation of living cells on a nutrient medium for experiments.

Centers for Disease Control and Prevention (CDC): Atlanta-based Federal agency under the Department of Health and Human Services; in charge of monitoring and controlling disease outbreaks in the U.S.

central nervous system: The brain, cranial nerves, and spinal cord; excludes the peripheral nerves.

cerebrospinal fluid: Clear fluid from the *meninges* that envelops the brain and spinal cord. Samples are drawn through lumbar punctures ("spinal taps"). Cloudy fluid indicates brain inflammation.

chronic wasting disease (CWD): A North American *prion* disease of elk and deer (mule deer in the west, white-tailed deer in the east). Animals typically lose weight and drool excessively.

CJD Surveillance Unit: Monitors all cases of Creutzfeldt-Jakob disease in the U.K. Based in Edinburgh, Scotland.

codon: A sequence of three *bases* specifying an *amino acid*. Also used to refer to the amino-acid position in a protein.

codon 129: The most critical area of the human *prion protein* gene. Either *methionine* or *valine* can be coded at this position.

Congo red: A dye used to stain *amyloid plaques*.

Creutzfeldt-Jakob disease (CJD): The most common *prion* disease in humans. It comes in inherited and sporadic forms. Compare with *variant Creutzfeldt-Jakob disease (vCJD)*.

dalton: A unit of mass used in biochemistry. It is defined as being equal to 1/12 the mass of a single atom of the most common isotope of carbon (carbon-12). Equivalent to atomic mass unit.

DEFRA: Department for Environment, Food and Rural Affairs. The British equivalent of the U.S. Department of Agriculture. It replaced the *Ministry of Agriculture, Fisheries, and Food* in 2001.

distal ileum: Lower part of the small intestine.

DNA: Deoxyribonucleic acid—a *nucleic acid* that carries the genetic information in the cell and is capable of self-replication and synthesis of *RNA*. It consists of two long chains of *nucleotides* twisted into a double helix and joined by hydrogen bonds between the complementary *bases* adenine and thymine or cytosine and guanine. The sequence of *nucleotides* determines individual hereditary characteristics.

doppel: A protein similar to the *prion protein* and whose overexpression can cause illness similar to a *prion* disease.

downers: Cattle that cannot stand on their own over a 24-hour period.

dura mater: The outermost membrane of the three-layer *meninges*, the thin covering over the brain.

dysarthria: Slurring of speech due to problems of muscle control.

encephalopathy: Any disease of the brain.

endemic area: In *chronic wasting disease*, it refers to the zone where the disease was first recognized. The patch of land is shared by Colorado, Wyoming, and Nebraska and encompasses approximately 15,000 square miles.

endosome: A compartment inside a cell that acts as a vehicle to move material around.

epidemiology: The statistical study of disease patterns, used to determine the cause and extent of illnesses.

epitope: A specific region on an antigen that binds to an *antibody*.

eradication zone: In *chronic wasting disease*, it refers to a 411-square mile area in southwestern Wisconsin, where, at the time of this writing, officials hope to kill off 25,000 deer.

fatal familial insomnia (FFI): An inherited *prion* disease of humans affecting the brain's sleep centers and thereby preventing the patient from falling asleep. Also called fatal insomnia to account for sporadic cases.

feline spongiform encephalopathy (FSE): A *prion* disease of cats.

Fore: A group of people in the Eastern Highlands of Papua New Guinea that once practiced cannibalism, which lead to the spreading of *kuru*.

formaldehyde: A disinfectant that, when mixed with water, makes formalin, which is used to preserve biological specimens.

14-3-3 protein: A type of protein that, when present in the *cerebrospinal fluid*, serves as an indicator of brain damage.

GAO: General Accounting Office, the investigative arm of the U.S. Congress.

Gerstmann-Sträussler-Scheinker syndrome (GSS): An inherited *prion* disease of humans primarily affecting motor functions.

gliosis: The proliferation of glial cells in the brain; a sign of neuronal damage.

glycosaminoglycans (GAG): A class of sugar molecules that line many cell membranes and help support cells. May be involved in the conversion of PrP^C to PrP^{Sc}.

greaves: Protein produced by rendering.

Harvard Risk Assessment: Evaluation of the *BSE* risk for the U.S., conducted by researchers from Harvard University and Tuskegee University and released in November 2001.

heterozygous: Having different copies of the same gene.

histopathology: The microscopic study of diseased tissue.

homozygous: Having two identical copies of same gene.

iatrogenic: Describing adverse conditions brought on by medical procedures ("physician borne").

immunoassay: Testing a sample for certain proteins or other compounds using various techniques, such as *antibody* binding and fluorescence.

Institute for Animal Health: The U.K. organization that investigates animal diseases. It has laboratories in Compton, Pirbright, and Edinburgh, where the lab is named the Neuropathogenesis Unit.

in vitro: "In the glass"; refers to test-tube studies.

in vivo: "In the body"; refers to human and animal studies.

incubation: Time between infection and the clinical manifestation (show of symptoms) of the disease.

index case: The first case in a specified group.

intracerebral inoculation: Injection into the brain.

intraperitoneal inoculation: Injection into the abdominal cavity.

kuru: An acquired *prion* disease spread by cannabilistic practices among the *Fore* people of New Guinea (now Papua New Guinea).

lysosome: Cellular compartment that breaks down material.

mad cow disease: *BSE.*

meat-and-bone meal (MBM): A protein supplement for livestock made by rendering slaughterhouse remains. It usually contains material from different species.

mechanically recovered meat (MRM): Meat extracted from a carcass using high-pressure equipment. Often used to make sausages, meat pies, and other processed foods.

methionine: An *amino acid* that plays a key role in *prion* diseases when occupying *codon 129.*

Ministry of Agriculture, Fisheries, and Food (MAFF): The British equivalent of the U.S. Department of Agriculture. Replaced by *DEFRA* in 2001.

meninges: The thin covering over the brain.

mutation: A change in the genetic code. In point mutations, a single *base* pair in *DNA* is altered, leading to an *amino acid* substitution. In insertional mutations, extra *amino acids* are added in.

myelin: Fatty substance that sheaths some neurons.

myoclonus: Brief, abnormal muscular jerkings.

neurodegenerative: Relating to the loss and damage of neurons.

neuropathology: Examination of brain tissue.

NIH: U.S. National Institutes of Health. Main campus is at Bethesda, Maryland.

nucleic acid: *DNA* or *RNA*.

nucleotide: The principal component of *nucleic acid*, consisting of a sugar and phosphate backbone attached to a *base*.

offal: Unwanted parts of a slaughtered animal, generally entrails and some internal organs.

Office International des Epizooties (OIE): Paris-based World Organization for Animal Health.

passaging: In *TSE* research, refers to the transmission of an infectious agent through successive experimental animals.

pathology: The study of diseases and the changes they produce in the body.

pentosan polysulfate: A drug for bladder infections that also shows promise as a treatment for *prion* diseases. It was injected directly into the brain of two *vCJD* patients in early 2003—without success.

prion: A *pro*teinaceous *in*fectious particle. Now widely defined as a protein that can pass on hereditary characteristics as *DNA* and *RNA* do and assume two shapes. In humans, those shapes correspond to a normal one (PrP^C) and a pathogenic one (PrP^{Sc}).

PRNP: The name of the human *prion protein* gene.

prion protein (PrP): Often used to refer to the normal, cellular form, PrP^C (sometimes called PrP-sen, for its sensitivity to enzyme digestion). The pathogenic, "*scrapie*" form is PrP^{Sc}, or PrP-res (for resistance to enzymes).

PrP 27-30: The indigestible core of PrP^{Sc}.

protease: An enzyme that breaks apart proteins.

proteinase K: A type of *protease*, used to digest *prion protein*.

proteosome: A cellular compartment that specifically breaks down proteins.

[PSI]: A yeast *prion*, representing an altered form of the yeast protein called Sup35.

quinacrine: An antimalarial drug used in experimental treatment of Creutzfeldt-Jakob disease.

recombinant: Refers to the intentional recombination of genes; genetic engineering

rendering: The process of converting slaughterhouse remains into different products, such as candles, soaps, and animal feed.

ribosome: Protein-making parts of the cell, where messenger *RNA* is decoded and *amino acids* are strung together.

RNA: Ribonucleic acid—a polymeric constituent of all living cells and many *viruses*, consisting of a long, usually single-stranded chain of alternating phosphate and ribose units with the *bases* adenine, guanine, cytosine, and uracil bonded to the ribose. The structure and base sequence of RNA are determinants of protein synthesis and the transmission of genetic information. RNA that delivers the *DNA* code to the *ribosome* is called messenger RNA; *amino acids* are brought to the ribosome by transfer RNA.

ruminant: An animal that chews its cud (partially digested food regurgitated from the rumen) and has a four-chambered stomach.

SAF: *Scrapie*-associated fibrils characteristic of *TSEs*. Also known as *prion* rods.

Scientific Steering Committee: A group of European scientists that provides scientific and technical advise to the European Commission.

scrapie: A *prion* disease of sheep, characterized by itching so intense that sheep scrape themselves raw.

Southwood Working Party: A U.K. panel in existence from 1988 to 1989 and led by Sir Richard Southwood to assess the threat of *BSE* and advise the government.

specified bovine offal (SBO): Cattle tissue banned from consumption.

spongiform encephalopathy: See *transmissible spongiform encephalopathy (TSE)*.

spongiosis: The formation of spongy holes in tissue.

Spongiform Encephalopathy Advisory Committee (SEAC): Formed in April 1990 to advise the U.K. on matters concerning *prion* diseases.

subclinical: Infected cases that have yet to show symptoms.

tallow: Fat produced by rendering.

transmissible mink encephalopathy (TME): A *prion* disease of mink.

transmissible spongiform encephalopathy (TSE): Any disease that produces microscopic holes in the brain, so that the tissue resembles a sponge, and that can spread to other individuals. Original name for *prion* diseases.

TSEAC: Transmissible Spongiform Encephalopathy Advisory Committee—a committee of scientists who provide advice and recommendations concerning prion diseases to the U.S. Food and Drug Administration.

[URE3]: A yeast *prion*; it represents an altered form of the yeast protein Ure2p.

valine: An *amino acid* that plays a critical role in *prion* diseases when occupying *codon 129*.

variant Creutzfeldt-Jakob disease (vCJD): The human form of *BSE*, distinct from sporadic and inherited *Creutzfeldt-Jakob disease*.

virino: Hypothetical disease agent that uses a host's own protein as its coat.

viroid: *RNA virus* lacking a protein coat; invades plants.

virus: Nonliving entity consisting of *DNA* or *RNA* surrounded by a protective protein coat.

WHO: World Health Organization, based in Geneva, Switzerland.

zoonosis: Any animal disease that can infect humans. (Adjective: zoonotic.)

For Further Information

Online Resources

Priondata.org:
http://www.priondata.org/
A site run in part by Stephen Dealler, a medical microbiologist who warned about the threat of BSE early on. It contains different levels of information and commentary, some for a fee.

The U.K. Creutzfeldt-Jakob Disease Surveillance Unit:
http://www.cjd.ed.ac.uk/
Information and links especially about vCJD.

U.K. Department of Health website for CJD:
http://www.doh.gov.uk/cjd/
Contains monthly CJD statistics in the U.K.

Chronic Wasting Disease Alliance:
http://www.cwd-info.org/
CWD information and tips for hunters.

World Health Organization:
http://www.who.int/health-topics/tse.htm

Centers for Disease Control and Prevention:
http://www.cdc.gov/ncidod/diseases/cjd/cjd.htm

Food and Drug Administration:
http://www.fda.gov/cber/bse/bse.htm
Includes information about blood donations.

Animal and Plant Health Inspection Service, USDA:
http://www.aphis.usda.gov/lpa/issues/issues.html
Information about CWD and BSE surveillance in the U.S.

SUPPORT AND HELP

Human BSE Foundation:
http://www.hbsef.org/
A U.K. organization, founded by Dave and Dorothy Churchill.

CJD Foundation:
http://cjdfoundation.org/

CJD Voice:
http://members.aol.com/larmstr853/cjdvoice/cjdvoice.htm

Books

Rhodes, Richard. 1998. *Deadly Feasts: The "Prion" Controversy and the Public's Health.* New York: Touchstone Books.

Transmissible spongiform encephalopathies from a mostly historical perspective. This book contains what will probably be the last media interviews of D. Carleton Gajdusek, before his legal troubles set in. The book also takes an "anti-Prusiner" tone and focuses on the viral thinking behind TSEs.

Rampton, Sheldon, and John C. Stauber. 1997. *Mad Cow USA: Could the Nightmare Happen Here?* Monroe, Maine: Common Courage Press.

A look at the cattle industry, seen through environmentalist and muck-raking perspectives. It argues that U.S. regulations did not do enough to prevent BSE, although some of the arguments are now dated. The book contains the last interviews of Richard Marsh before his death.

Index

Page numbers in *italics* refer to illustrations.